TIME,
CONTINUITY,
AND
INDETERMINACY

TIME, CONTINUITY, AND INDETERMINACY

A Pragmatic Engagement with Contemporary Perspectives

SANDRA B. ROSENTHAL

State University
of New York
Press

Published by
State University of New York Press, Albany

Production by Susan Geraghty
Marketing by Anne Valentine

Printed in the United States of America

For information, address State University of New York Press, State University Plaza, Albany, N.Y., 12246

Library of Congress Cataloging-in-Publication Data

Rosenthal, Sandra B.
Time, continuity, and indeterminacy : a pragmatic engagement with contemporary perspectives / Sandra B. Rosenthal.
p. cm.
Includes bibliographical references and index.
ISBN 0-7914-4493-7 (alk. paper). — ISBN 0-7914-4494-5 (pbk. : alk. paper)
1. Time. I. Title.
BD638.R68 2000
115—dc 21 99-15654
CIP

10 9 8 7 6 5 4 3 2 1

"Yesterday's the past, tomorrow's the future, but today is a GIFT. That's why it's called the present."

For
Steve and Sandy
and
Leslie and Scott

Who help make my present so precious

CONTENTS

ACKNOWLEDGMENTS

I wish first and foremost to acknowledge my indebtedness and gratitude to two very special individuals, without whom this book would never have been completed. During the numerous times I put this manuscript away for months on end in a sense of frustration, my husband, Rogene, kept gently reminding me that I had something I very much wanted to say and I had better get back to the job of figuring out how to say it. When I finally thought I had "thrown in the sponge" for good, John Lachs suggested I send the manuscript along to him and, as he has done so often throughout my career, offered encouragement, enthusiasm, and insightful suggestions.

Additionally, Peter Hare and David Weissman read various parts of an initial draft some years ago, and their criticisms led me to a needed pause for reflection on my method of philosophizing, its strengths and its limitations.

My thanks also go to Loyola University for a yearlong sabbatical that, though granted for other projects, provided me with the opportunity to let the ideas for this one begin to germinate.

Parts of the present work appeared previously in the following journals. They are reprinted here with the kind permission of the publishers.

Discreteness, Continuity, and the Fate of Time: Contemporary Perspectives," *International Philosophical Quarterly* 37 (1997). Copyright 1997 by Fordham University.

"Contemporary Metaphysics and the Issue of Time: Rethinking the Great Divide," *International Philosophical Quarterly* 39 (1999). Copyright 1999 by Fordham University.

"Temporality and the A Priori: Lewis and Heidegger," *International Studies in Philosophy* 27 (1996). Copyright 1996 by Binghamton University.

"Continuity, Contingency, and Time: The Divergent Intuitions of Whitehead and Pragmatism," *Transactions of the Charles S. Peirce Soci-*

ety: A Quarterly Journal in American Philosophy 32 (1996). Copyright 1996 by the University at Buffalo, State University of New York.

"Self, Community, and Time: A Shared Sociality," *The Review of Metaphysics* 50 (1996). Copyright 1996 by the Catholic University of America.

"Scientific Method and the Return to Foundations," *The Journal of Speculative Philosophy* 2 (1988): 192–205. Copyright 1988 by the Pennsylvania State University. Reproduced by permission of the Pennsylvania State University Press.

"Contemporary Process Metaphysics and Diverse Intuitions of Time: Can the Gap Be Bridged?" *The Journal of Speculative Philosophy* 12 (1998): 271–288. Copyright 1999 by the Pennsylvania State University. Reproduced by permission of the Pennsylvania State University Press.

"The Family Circus," reprinted with permission of King Features Syndicate.

INTRODUCTION

There are many ways to approach philosophical positions in order to understand their underlying affinities and differences, and differing approaches may at times result in quite differing alignments. Approaching them through the understandings of time that their respective positions harbor is one way, and a very fundamental way, of offering illumination, perhaps cutting to the core of a position more radically than is possible through the use of other pathways. The present work, in focusing on the issue of temporality, will first explore the frameworks of some of the philosophers who have in large part shaped the contours of the contemporary philosophical landscape. In so doing, it will remap the terrain, often finding similarities where differences—some quite radical—are generally accepted, and finding differences where similarities are generally accepted.

The present endeavor does not intend to deny that there are blatant, significant, indeed crucial differences between positions that will here be viewed as similar, or broad and deep affinities between positions that will be viewed as different. But it holds that at a very fundamental level the positions discussed in the following pages are either joined at the hips or separated at birth, so to speak, through certain temporal assumptions, often implicit and at times explicitly denied, that underlie their various philosophical claims and structure their various problematics. Exploring these temporal assumptions can hopefully cast a novel light upon the contemporary landscape in a way that can prove fruitful in providing an opportunity for an ongoing and deepened constructive dialogue about fundamental issues of philosophy in general and time in particular that need to be discussed and that are too often not only housed by but hidden within the structure of various frameworks.

The approach to the analyses and critiques of the various positions discussed in the following pages is blatantly pragmatic and is guided by the fundamental intuitions of time that pervade the pragmatic perspective. A pragmatic reconstruction of time and related issues that these intuitions involve will be explored and developed in the concluding part, its significance best coming into focus from the backdrop of the various temporal assumptions of the philosophers discussed earlier. Some readers, operating within different intuitions of time, may think the entire work suffers from the "Humpty-Dumpty" syndrome, a syndrome based on the intuition that discrete units cannot adequately be brought together.[1] They will hold that discreteness does not pose the problem this work claims. Differing intuitions of time are both deeply embedded and pervasive within a position and color in many ways not only its problematics but also the solutions considered adequate. Because these intuitions run so broadly and deeply and are at times operative in either an implicit or unintentional way, it is all the more important to elicit them for examination as a way of viewing the contours of the philosophic terrain.

The issue of time of course involves the notions of past, present, and future. And related to these are the concepts of potentiality, possibility, and actuality. Moreover, undergirding them all is the question of the relation—or lack thereof—between discreteness and continuity, and inextricably interwoven with the issue of discreteness and continuity is that of fixity and indeterminacy. The position taken on these issues is in turn crucial for the various understandings of the ontological or epistemic or phenomenological status of our sense of temporal passage.

From the pragmatic perspective, time is not a container for, or something "in itself" apart from, the temporal characteristics of what and how we experience. The features of temporality are the features of a temporal universe and its pervasive ways of behaving, ways that permeate the universe in general and various emergent realities within the universe, including human activity and human awareness of the universe in which it is embedded. Thus in some chapters this work will not focus directly on the nature of time but on the implications of a particular stance on time for various issues, some of which are usually considered far removed from concerns with temporality, such as the understanding of self, community, scientific method, and the a priori.

Perhaps nothing has more dominated the history of Western philosophy, either explicitly or implicitly, than the serial view of time, the view of time as discrete, and this focus on discreteness is even more pervasive than may at first appear. More often than not if there is an attempt to attain a continuity in temporal passage it begins with the temporal foundationalism of atomic moments, and from there some way must be found to bring forth the philosophic glue. If one does begin with continuity, discreteness again tends to raise it head, for the continuum is understood as offering a mathematical divisibility into denumerable parts, as a result yielding some of the same paradoxes.

As far back as antiquity Zeno exposed the paradoxes arising from such views. Aristotle, in rejecting Zeno's paradoxes, argued that time is not actually but potentially divided into an infinite number of parts, while the actual reality of times consists of stretches and intervals. Yet a truncation of parts slips into his claim that the point of time that divides earlier from later must be held always to belong to the later to avoid the case that the same moment both is and is not, and that something is not at the moment when it has become. Similarly Augustine's famous discussion of time and the paradoxes involved presupposes the necessity of a discrete isolatable present.

The modern period of philosophy, intertwined as it was with the period of modern physics, the age of the Newtonian worldview, both incorporated and reinforced the atomistic view of time coming up from the ancient and medieval periods. From the direction of philosophy and science alike, the serial view of time and the discreteness it involved took on added authority. The discrete view of time is to be found in philosophers as diverse as Leibnitz, Descartes, Hume, and Kant. It is involved in the radical skepticism of Hume, while Descartes finds in the "obvious" discreteness of time the need for God's ongoing existence in order to create us anew at every moment, a God that ultimately saves him from skepticism. The need to provide some sort of a metaphysical unification of the members of the discrete series by moving to something beyond it leads most frequently to the move to God, but is found operating in a different way in Kant's use of the transcendental ego to unify discrete data.

While some of the contemporary discussions of time can be traced back to their roots in antiquity and others to the rise of modern physics, the rise of contemporary physics introduced into the issues of time an added factor in the need to ground the arrow of time, now not only in the face of the reversibility of scientific laws but in the face of temporal relativity. The scientific critique of the forms of space and time that emerge from non-Euclidean geometry and relativity physics leads to the paradoxes involved in a radical plurality of times. The idea that my present is simultaneous with the future of a different observer sufficiently distant from me destroys the very meaning of past, present, and future. Thus Einstein, in insisting that it is to science alone that we must go for the truth about time, and hence that lived time with its experienced sense of past, present, and future is a "persistent illusion," nonetheless recognized that scientific descriptions cannot satisfy our human needs; that there is something about our sense of time that is "just outside the realm of science."[2] The truth concerning time in science is contained in mathematical calculation, but our lived sense of time stubbornly resists attempts to reduce it to the content of mathematical systems.

Meanwhile, the problem of temporal discreteness remains obstinately present. The scientific understanding of a space-time continuum can be somewhat misleading, for while space and time become continuous with each other, this entire concept has led to a spatialization of time in which time can be understood as a label identifying each slice in a stack of spaces, with each slice in the stack representing the whole universe of space at a different time. The whole space-time block can be sliced up into a stack of "time-slices" in many different ways, from many different angles. Further, the connection between matter and space-time geometry indicates that time can be defined internally by some geometrical property, like the curvature of each slice. The four-dimensional manifold is a mathematical device for interpreting physical measurements. It must assume that the time axis is at right angles to the three explicitly spatial dimensions, and it can do so only in the "configuration space" of mathematics. The spatialization of time in physics is a laudable stance for the advancement of physics, but, as a mathematical device, it again abstracts from temporal passage, and the entire metaphor of slices of stacks of space smacks of a transformed discreteness of temporal bits, which is inevitable given the mathematical apparatus upon which it depends.

It is of course precisely the relation between scientific time, the temporality of human existence, and concrete cosmic or "metaphysical" time that is often at issue in contemporary philosophical discussions. Whatever the answer a particular philosopher provides, however, the importance of undergirding the arrow of time in some sense and of specifying in some way the relation of the three types of time has taken on added urgency. And underlying this endeavor lurks the ongoing problem of discreteness or continuity and the related issues this involves.

The first section of this work focuses on the pervasiveness of the understanding of time as discrete. Chapter 1 concentrates on two giant shapers of contemporary speculative philosophy, A. N. Whitehead and Paul Weiss, and the nature of the relationship between Whitehead's process metaphysics and Weiss's substance metaphysics as viewed from the perspective of their temporal assumptions. Chapter 2 turns to two dominant figures in the growing concern with the demise of time, Jacques Derrida and J. M. E. McTaggert. While Derrida's deconstructive position grows out of the continental tradition, and McTaggart's arguments for the unreality of time is emmeshed within the Anglo-analytic tradition, their respective assumptions concerning time provide a common denominator by which to explore their relationship.

Part II focuses on Heidegger's rejection of temporal discreteness, showing the way his temporal concerns can be both illuminated and expanded by exploring an underlying temporal kinship with pragmatic philosophy. Chapter 3 recasts and expands his concerns with lived temporality and his halting tendencies to move beyond this limited focus. Chapters 4 and 5 of this part examine in turn the way common intuitions concerning temporal issues in Heideggerian and pragmatic philosophy allow for their shared understandings of the nature and function of scientific method and the a priori element in experience. Here their respective expressions of a broadly based reformulation of these issues can each help illuminate the other, and this mutual illumination can in turn help highlight the far ranging significance of a shared understanding of time.

The final part of the book turns more directly to the development of a pragmatic framework for temporality and related issues. Chapter 6 delves into the temporal problems and issues that pervade Mead's understanding of the self, offering in the process a reconstruction of his position that is more fully in line with his pragmatic insights concerning temporality in general. The following chapter presents an overview of the way the issue of temporality enters into contemporary understandings of community shaped by Rawls, MacIntyre, and Rorty, and then examines a radically different pragmatic perspective rooted in the temporal dynamics of selfhood developed in the previous chapter. The remaining two chapters explore, in more broadly based and systematic fashion, a pragmatic understanding of the temporal features of our lived experience and the universe at large, thereby developing the general context that in fact set the contours within which all other chapters took their shape. These two chapters bring to light more fully the way the pragmatic understanding of temporality involves a radical reconstruction of traditional ways of understanding and interrelating the key issues of time—discreteness and continuity; fixity and indeterminacy; potentiality, actuality, and possibility; past, present, and future.

As a final note, it should be pointed out that the pragmatic perspective being taken throughout this work is not intended to be the product of a historical analysis of the doctrines of the classical American pragmatists. Rather, discussions of the pragmatic position will at various times, and in varying degrees in various chapters, utilize unchanged certain doctrines of one or several of the pragmatists, modify certain claims, develop what lies incipient in their thought, creatively synthesize and extend the partial claims of several or all of them, and develop certain features that are inspired by, but not to be found in, the collective corpus of their work. These features will at times differ from and in fact may contradict some of their particular relevant doctrines. However, at no time will "the pragmatic position" taken in this work depart from what this author sees as the spirit or general intent of their thinking.[3]

PART I

The Pervasiveness of the Discrete

CHAPTER 1

Whitehead, Weiss, and Speculative Metaphysics of Time

One of the "great divides" in contemporary metaphysical debates is the issue of substance versus process. In metaphysical circles, Whitehead is by and large considered *the* process philosopher, and contemporary process philosophers such as Robert Neville,[1] among others, are distinctively Whiteheadian in their approach. The most systematic substance philosopher of contemporary times, Paul Weiss, engages Whiteheadian process as his protagonist, and his position can be viewed as a response to perceived problems in Whiteheadian process. The present analysis, in exploring this speculative terrain with the tool of temporality, will restructure the "great divide," placing all of the above on one side, and a pragmatic process metaphysics of activity on the other. For, while Whiteheadian process and pragmatic process are similar in much of what they attempting to do, they ultimately are at opposite poles in their manner of doing so because of their different intuitions of time and the related issues of continuity and discreteness, potentiality and actuality, indeterminacy and fixity, which structure and pervade their respective attempts. On the other hand, the intuitions in these areas that guide Weiss's development of a contemporary substance philosophy in explicit opposition to Whiteheadian process are in fact remarkably similar to those that structure the latter. What ultimately places the pragmatic perspective in opposition to both of these movements is the fact that its fundamental intuitions in these areas lead to its riddance of all remnants of the fully discrete, the fully actual, the fully determinate, the fully fixed. And it is through this lens that the exploration of "the other side of the great divide" will proceed.

The enormous contribution to philosophy of the philosophers to be discussed in the present chapter is beyond dispute. That this chapter concentrates on what it sees as crucial problem areas is not intended in any way to deny the magnitude of their contributions but to recognize that precisely because of this it is important to pinpoint the fundamental divide between them and pragmatic process philosophers and the way in which this divide is more basic than that which separates them as process and substance philosophers. With the above points in mind, the discussion will turn to the relevant features of Whiteheadian process.

There are three interrelated influences on the direction of Whitehead's thinking that will be most relevant to the issue of time in the following discussion:

1. The problems he finds with Bergson's understanding of time
2. His concern to avoid the bifurcations of the modern period by finding a basis in experience for the findings of science in general and scientific time in particular, at the same time eliminating the notion of time as an absolute container
3. The problem of maintaining the arrow of time in the face of the reversibility of the laws of science and the multiple time frames of relativity theory.

Whitehead, like Bergson, objected to the priority of the physicists' claim to truth, including the spatialization of time in physics, and turned to process philosophy as a remedy. However, he rejected Bergson's claim that there is an indivisible "pure time" or "real duration" that can be grasped by intuition or a passive sympathetic subjective attunement with the flux of pure becoming. For Bergson, there was, in Whitehead's words, pure, unadulterated inner continuity (duration), continuity which was neither unity nor multiplicity."[2] Whitehead objected to Bergson's conflation of change and becoming, for there could be no real change in a purely continuous flow. Neither time nor motion can be clarified by a duration in which multiplicity is abolished; if there is no multiplicity there can be no time. Rather, according to Whitehead, the continuity of the stream of experience is a surface feature prominent for consciousness, but the underlying reality is a succession of atomic experiences, "actual entities or actual occasions of experience." In Whitehead's position, the concrete and inclusive realities, from which all other realities are derivative by abstraction, are actual entities. They are the ultimate units of the universe and are not themselves divisible into units. Actual occasions or actual entities become themselves in momentary processes of concrescence—processes of becoming concrete, actual, or determinate; and, upon fully achieving this status they at once perish. The moment of becoming of an actual entity is also its moment of perishing.

Whitehead's emphasis on atomicity is in part a reaction to the Bergsonian pure duration, but his concern to prevent the bifurcation of nature also leads him to the atomicity to which he is committed. The modern worldview understanding of Newtonian physics led to the bifurcation between the world of sense and the world of science, and this in turn led to insoluble epistemological dilemmas in modern philosophy, such as the bifurcation of object and subject or observer. The latter occurred because the Newtonian bifurcation was understood as a metaphysical bifurcation: physics became the metaphysics of nature. Einstein likewise made a bifurcation between nature as sensed and nature as an object of science, reifying this into a metaphysical distinction in which the real is what is contained within the world of science. Whitehead saw the need to avoid the epistemological problems by rejecting the metaphysical bifurcation.

In rejecting the bifurcation of nature, Whitehead rejected the distinction between nature as sensed and nature as designated by scientific theory. He did not distinguish between sense objects and scientific objects on the basis of Newtonian and Einsteinian bifurcations that held that colors and sounds are sensed data and electrons and protons are unobserved postulated entities too small for the senses to detect. Rather, he held that both atoms and sense data are immediately sensed adjectives of immediately sensed events; the difference lies in the fact that adjectives that are colors and sounds depend upon the relation between the event that is their sensed locus and the percipient event that is the sensed observed, while the adjectives that are scientific objects such as electrons and protons are more persistent and are a function only of the events that are their sensed locus.[3] An electron is a sensed adjective qualifying all the event particles in a "historical route."[4] Scientific concepts are derivable from what is immediately sensed by abstraction.

Whitehead wanted both to show that beginning with absolute space and time as containers is an example of the fallacy of misplaced concreteness, that the uniformity of space-time is an abstraction from concrete experience, and to explain how the time of physics can be derived by extensive abstraction from a multiplicity of concrescences or units of becoming and transitions between them. Measurable aspects of nature, such as points of space and instants of time, are abstracted from the fundamental characteristics of nature. The point time of physics is achieved by extensive abstraction from the duration of atomic units of time.[5] Extensive abstraction is applied, not first to durations and then to space alone, but first to regions of space-time, so that the definitions of spatial entities are not dependent on the prior defining of temporal entities, but can be introduced side by side with them. As Whitehead affirms, "it needs very little reflection to convince us that a point in time is no direct deliverance of experience, but it is obtained by extensive abstraction. We live in durations, and not in points."[6] But the discrete point times of physics are obtained by extensive abstraction from discrete units of temporal duration. Indeed, although Whitehead attempts to present a grandly unified view of humans and the reality within which they are embedded, it is with the conclusions of science that this quest begins. His development can be traced by understanding it as providing evolving solutions to the dual problems of irreducibly atomic physical occasions: bringing them together and fusing them with life.[7]

The need to preserve the arrow of time again leads to atomicity, this time to guarantee the absolute fixity of the past. This involves other "fixities" as well. While it may at first seem that the issue of substance versus process philosophy is the issue of fixity and definiteness versus change and indeterminacy, yet the philosophical search for the security of fixity, the fully definite, as well as the atemporal, has survived surprisingly well in Whitehead's philosophy. Actual occasions, while a process of becoming, have their own type of "fixity," for they do not change, do not alter while maintaining an identity. Rather what changes is a serial "society" of actual entities. Persistence through change is not the result of any underlying enduring actual occasion but the retaining of defining characteristics as other characteristics are replaced in an ongoing succession of actual occasions. In our experience we normally discriminate not a single actual entity but a nexus of them united by their prehensions or "concrete facts of relatedness."

What survives the temporal passage, the ongoing succession of actual occasions in their atomic moments of self-realization and perishing, is the potentiality that "survives" and is prehended, in this way entering into further actual occasions. The "objective immortality" of past actual occasions lies in their potential for prehension by later occasions. The ongoing fixity of the past through the perishing of actual entities that have prehended it allows for the arrow of time and the preservation of the tensed distinctions of past, present, and future. Time, then, is tied to the succession of atomic actual occasions that "perish" to become a fully fixed, unchanging past. This is essential to time in two respects. First, the sense of the distinction between past and present requires that something must perish. The arrow of time requires this perishing. As Whitehead stresses, "the irreversibility of time depends on this character."[8] Second, the absolute definiteness of perishing actual occasions ensures the absolute fixity of the past, which he also views as a requirement for the arrow of time. Apart from the succession of actual entities, time does not exist, for this succession constitutes, in Whitehead's words, "the becoming of time."[9] Actual entities are not "in time," but rather in their atomic succession give rise to time.

This ongoing succession of perishing events gives rise to time, and thus time is constituted by discrete stages or, in Whitehead's terms, epochs. In its fundamental character, then, time is a discrete succession of epochs or arrests, each being the duration required for the emergence of a prehensive unity as a single fact,[10] and each prehensive unity is itself an indivisible epoch having no internal temporal phases. This constitutes Whitehead's epochal theory of time, which, as Victor Lowe has well observed, might better have been called the "atomic theory of process."[11] Time, like actuality, is for Whitehead "incurably atomic."

Because of the absolute fixity of the past, Whitehead's system requires the fixity of eternal objects to account for novelty. For, if the past provided everything for the present, then nothing new could appear; the apparently new would be merely a reconfiguration of the old. Since the past is fixed and final, novelty is possible only by the intervention of something eternal or atemporal; the future exists as possible in the form of inert, fixed, objective real possibilities. The ingression of eternal objects into events is Whitehead's way of getting contingency from a fixed, fully determinate past that cannot itself adequately allow for contingency and emergence. Every actual occasion introduces novelty into the world by prehending and newly integrating what the past sends to it, and it does this via a "lure" directed to the future through its prehension of eternal objects.[12]

Eternal objects are pure potentials or forms of definiteness, timeless potentials for every process limited only by the logical demands of being mutually consistent or alternative. These are housed in the eternal primordial mind of God as the one eternal actual entity. The potentiality for an actual entity, however, is never pure, a function of the eternal, but involves as well what is already actual. As Whitehead states, "A new creation has to arise from the actual world as much as from pure potentiality; it arises from the total universe and not solely from its mere abstract elements."[13] The past places limitations on the ingression of eternal objects into actual occasions and is incorporated into the present through prehensions as the way in which an occasion can include past actual entities and eternal objects as part of its own essence.[14] While limitations are placed on the ingression of eternal objects by the past, yet such limitations neither "create nor destroy" them, for there must be an excess of unrealized potentialities provided by eternal objects if one is to avoid an understanding of the universe as static and ultimately monistic.

It has been seen that actual occasions are the atomistic building blocks of time; they give rise to temporal extension but are not themselves temporally extended in their becoming. As Whitehead states, "genetic passage from phase to phase is not in physical time." Rather, "physical time expresses some features of the growth, but not the growth of the features."[15] Yet Whitehead's genetic analysis of concrescence or the process by which actual entities achieve definiteness sounds decidedly tensed, for in the process of concrescence there are "phases in which new prehensions arise by integration of prehensions in antecedent phases."[16] As Donald Sherburne observes, " It seems strange to speak of one phase of concrescence as before or after another when the passage from phase to phase is 'prior to' physical time. This is a difficulty."[17] Sherburne and William Christian[18] agree that while this priority cannot be temporal priority, neither is it logical priority, part-whole priority, or dialectical priority. Rather, it must be a priority sui generis. Sherburne concludes in perhaps understated fashion that "it may seem to many that this suggestion is ad hoc and unsatisfactory. But a better suggestion is hard to find. This difficulty does not admit of easy solution, and Whitehead scholars increasingly are going to have to puzzle over it, for the sequence of phases of concrescence is too central to the system to be abandoned, and yet difficult to incorporate into the system."[19]

This problem has been dealt with by later scholars of Whitehead's work in various ways. One line of argument holds that the process of becoming internal to an occasion "is not a move from earlier to later stages but from an indefiniteness of possibility to a definite actual arrangement of features."[20] This internal move from possibility to actuality is difficult to understand unless one understands it in terms of a change from one condition to another, and the critic recognizes this, for he goes on to assert that change *does* take place in the becoming of a present occasion.[21] Hence in trying to handle the puzzle, not only becoming but also change comes to be understood in tenseless terms; yet change seems to be that which is most associated with temporal movement. This leads Whitehead to emphatically distinguish between becoming as internal to actual occasions that do not change and change as involving the temporal succession of perishing actual occasions.

Attempts have been made to deal with this problem in a somewhat different fashion by claiming that the temporal quantum associated with an occasion is created all at once.[22] But it is perhaps questionable if the concept of "all at once" can be understood apart from time determinations. Rather, "all at once" seems to mirror the notion of the instantaneous "now," which in turn gains its meaning in terms of time. Further, since the moment of creation is the moment of perishing, creation "all at once" does not seem to allow for any durational span to actual entities. As a result, for example, the claim arises that Whitehead's position here is one of "instantaneous becoming."[23] That actual entities are not meant to be characterized in this latter way seems clear, for Whitehead holds that in the endless succession of epochs or arrests, each arrest is the epochal duration required for the emergence of a prehensive unity.[24] The attempt to deal with the problem of the tenseless nature of actual occasions while maintaining their durational span has led to the rejection of "the common assumption that if *a* is earlier than *b*, *a* must lie in the past of *b*. . . . Though each occasion comes into being "all at once," there is no instantaneous becoming."[25] Yet, as many have argued, and it would seem rightly so, relations of earlier and later presuppose the tensed relations of past, present, and future.

The various above discussions seem a determined attempt to deal with what is inextricably temporal in tenseless terms, introducing by the back door what was kicked out of the front. This problem of the process of becoming of atomic units whose succession gives rise to time but that are, in their momentary processes of becoming concrete, themselves tenseless, seems indeed to be a puzzle not easily solved. Dynamic becoming and duration invoke the notions of past, present, and future, of temporal flow and, dependent upon these, of the change involved in the actualization of possibilities. Yet actual occasions, though their succession gives rise to time, are themselves neither temporally flowing nor changing, and are in time only after they have perished. The above attempts to account for the process of the realization of actual occasions offer variously an all at once, an earlier and later, the actualization of potentiality, and change, each of which is truncated from the temporal notions of past, present, and future that give it meaning. These problems emerge because Whitehead makes a wedge between the tensed time of the process of succession of discrete units and the seemingly tenseless duration of becoming or the process of concrescence.

Whitehead's focus on temporal atomicity as a reaction to the inadequacy of the Bergsonian duration in accounting for either time or change is clearly evident here. Time, for Whitehead, is not "another continuous process."[26] Put in terms of discreteness and continuity, the durational becoming of an actual entity must be tenseless precisely because it is a continuous process, while the coming to be of tensed temporal modes requires a succession of ontologically discrete units. In avoiding the "seamless Bergsonian duration" the process of creative becoming is itself taken out of time and becomes instead a building block of time. From the perspective of Whitehead's reaction to the inadequacy of time as an absolute container, it can be said that actual entities are not "in time" but rather in their atomic succession give rise to time.

A further discussion of the issues emerging in the above analysis requires a turn to relevant features of Whitehead's concept of the extensive continuum that provides the continuity of potentiality rather than the atomicity of actuality. For Whitehead holds that "Realization is the becoming of time *in the field of extension*,"[27] and extension "is the complex of events qua their potentiality."[28] Further, "The reality of the future is bound up with the reality of this continuum. It is the reality of what is potential."[29] While actuality, for Whitehead is "incurably atomic," potentialities provide continuity. Temporalization itself is the ongoing process of atomization in the field of potentiality or the extensive continuum.

The extensive continuum in its most general sense is the widest, least determinate environment in which our present spatiotemporal cosmic epoch is set. It is the first determination of real potentiality as opposed to the pure potentiality or abstract possibility of eternal objects. As such, it is generally accepted that it has the ontological status of a derivative abstraction from actuality. As Whitehead states:

> This extensive continuum expresses the solidarity of all possible standpoints throughout the whole process of the world. It is not a fact prior to the world; it is the first determination of order—that is, of real potentiality—arising out of the general character of the world.
>
> This extensive continuum is "real" because it expresses a fact derived from the actual world and concerning the contemporary actual world."[30]

In what follows, the focus on the issue of temporal atomicity in Whitehead's position will leave the problem of the tenseless nature of the becoming of actual occasions and move to the broad problem of the relation of continuity and discreteness in general in Whitehead's position, the problem of the prehension or mutual emanance or connectedness of ontologically discrete actual occasions. The depth of this dilemma can perhaps best be illustrated through two sustained attempts to interpret and reconstruct Whitehead's position, the first from the direction of continuity, and the second from the direction of discreteness.

The implicit impetus for the first reconstruction stemming from a focus on the extensive continuum[31] seems to come from an implicit awareness of the problem involved in the claim that a continuum is derivative from a succession of discrete units. While recognizing that there is great textual justification for the accepted view that extension is a derivative abstraction from actuality, Jorge Nobo argues that without interpreting the extensive continuum as a metaphysical principle of Whitehead's philosophy there is difficulty making genuine sense of the mutual immanence of discrete actualities. In short, there can be no adequate understanding of solidarity or continuity or connectedness unless the extensive continuum is given an ultimate metaphysical status as opposed to a status as derivative from the actual world.

Thus, Nobo distinguishes two senses of the extensive continuum, one derivative and one metaphysically ultimate. The extensive continuum in the first sense is contingent and derivative from the ongoing succession of actual entities. The extensive continuum in its most general sense is necessary and metaphysically prior to it. It is an eternal "receptacle" within which the creative advance of actuality occurs. It does not itself become, but rather all actual occasions become within it.[32]

Nobo seems to be making a distinction between two senses of extensive continuum that can be understood as analogous to the two senses of potentiality, eternal potentiality or possibility and "real" potentiality as conditioned by what is provided by the actual world. He is of course doing more than explicating Whitehead; he is offering a reconstruction of Whitehead that seems to emerge from the problem of discreteness and continuity, though he never pinpoints it in this way. Perhaps the issue he is addressing can best be encapsulated by Whitehead's references to the "becoming of continuity."[33] As Whitehead states this in terms of time, "The temporal quanta do form a continuum, but the creation of this continuum is not a continuous becoming."[34] Nobo's implicit insight here is in recognizing a problem with the creation of continuity by noncontinuous, ontologically discrete units.

In discussing the required aboriginal properties of the continuum as metaphysical, he finds that Whitehead's understanding of continuity as unbounded and as indefinitely divisible though undivided, while adequate for dealing with mutual immanence, is not adequate to explain "not just mere immanence but the *mutual immanence of actual entities having non-overlapping extensive standpoints*," properties that can account for the connectedness of *discrete occasions*. Moreover, "no explicit consideration of the properties in question is to be found in Whitehead's works."[35]

Nobo constructs a feature for the extensive continuum that allows for the discreteness of actual occasions by extending the dividedness of actual entities from each other to the dividedness of extensive regions from each other. "Hence, proper extensive regions cannot include each other; they cannot even partially overlap so as to have common parts; they are, in short, mutually external. . . . But notice that, as used here, the 'separateness' of proper regions does not preclude their possible contiguity. Any two proper regions are separate even if they are contiguous."[36] And, while the extensive continuum is divisible, because of its separative character it is "actually divisible only into discrete or separate regions."[37] This separativeness of the extensive continuum, then, provides for a "metaphysical imposition of discreteness upon all actual entities."[38]

Once the extensive continuum has been made primary, then absolute discreteness can be saved only by giving up continuity in favor of a contiguity of discrete parts. In saving ontologically discrete units in the face of a now metaphysically prior continuity, Nobo has in fact again sacrificed continuity for the sake of discreteness, as a result negating an insight that seems to have been a strong though implicit impetus for his entire project. For a contiguous series of ultimately divided parts yields at best a pseudocontinuity.[39] Whitehead's lack of inclusion of separatist properties for the extensive continuum or the complex of events qua their potentialities seems far from an oversight. The above attempted reconstruction is indicative of the need in a Whiteheadian orientation to have an overburdened eternal to handle difficulties in bringing together the various ingredients of the temporal. Even a further over-burdening of the eternal, however, does not solve the problem of establishing a continuity of ontologically discrete units but rather reintroduces the problem from a different direction.

The problem of the relation of discreteness and continuity in Whitehead's position, this time coming from the direction of discreteness, can be further brought to light by briefly turning to another attempted reconstruction and the debate it has engendered. In attempting to portray a naturalized version of Whitehead's metaphysics of nature that excludes God, Donald Sherburne suggests a modification of the position that would allow immediately past, contiguous occasions to be causally efficacious, hence not requiring God and eternal objects. He puts forth the possibility that at the instant in which one occasion completes, and the next initiates, its becoming, the first can be both becoming and causing at the same time.[40]

To this Lewis Ford objects that this interpretation implies that two occasions must share a common interface, while Whitehead's theory of the external relation of discrete actual occasions as opposed to their internal relatedness via prehensions indicates that two contiguous acts of becoming are so close as to allow no point in between, but share no common region whatever. As Ford continues, it is not understandable that one occasion can continue to persist in the second as causal factor in its final satisfaction, if it instantly ceases to be at the end of concrescence.[41]

Sherburne's proposal, though very different from that of Nobo's, is similar in being a reconstruction that must mitigate the radical atomicity of actual entities in Whitehead's position in order to make sense of connectedness. Like Nobo, the answer is found in contiguity. While Nobo ultimately transforms a continuum into contiguous units in order to ground discreteness, Sherburne transforms discrete units into a contiguous series to avoid the problems of ontological gaps. In so doing, the former loses true continuity, while the latter does not reach it.

The above fundamental problems of temporality, continuity, and discreteness pervade all aspects of Whitehead's position. This can be exemplified by turning to his well-known attack on Hume's skepticism concerning causality, in which he argues that Hume mistakenly limited experience to the mode of presentational immediacy and ignored the more primordial experience of causal efficacy. According to Hume, our supposed sense of causal efficacy is merely our sense of our own habit of expectation. But, as Whitehead continues, Hume has in fact confused a habit of feeling one thing following another with the feeling of habit. If Hume cannot allow for the experience of cause, he cannot allow for the experience of habit.

Whitehead's criticism of Hume's lack of concern with experience in the mode of causal efficacy is well taken. But Hume's problem rests on something even more fundamental, and that is his understanding of time as discrete. It is impossible, given his theory of experience and the understanding of time in which it is rooted, to experience a causal connection because the ultimate components of time, and hence of temporal experience, are discrete. The internal relatedness of prehensions, for Whitehead, provides the continuity that allows for the experience of causal efficacy. But the problems raised earlier of interweaving continuity with ontologically discrete units are applicable here. Hume's skepticism shows, among other things, that contiguity in space and time, representing the proximity of discrete units, is inadequate. True, Hume's units of time are not durations of becoming, yet Whiteheadian durations of becoming are nonetheless like Hume's moments in being atomic building blocks, and the problem of relating metaphysically ultimate discretes, discussed above, seems to render ultimately inadequate Whitehead's attempt to account for the real sense of causal efficacy in experience that he so aptly and beautifully highlights. The solution to Hume's problem requires not just a rejection of atomic knife-edged moments but also a rejection of all vestiges of the ontological ultimacy of atomicity.

In Whitehead's position, potentiality and actuality are fundamentally isolated, for becoming actual is a process of achieving definiteness, and the moment of attaining full actuality and perishing is at once the attaining of absolute definiteness. There can be no residue of potentiality in the fully actual or it would not be fully actual, fully definite. Though potentiality "arises" from actuality, and actuality is a concrescence of potentiality, yet actuality and potentiality are strangely separated from each other, as are discreteness and continuity, present, past, and future, contingency and fixity.

Whitehead is concerned throughout with a temporalistic understanding of the immanence of the past in the present, but an understanding that begins with a chasm between the two that is difficult, perhaps impossible, to bridge. Thus the relation of past to present is a perpetual problem in the Whiteheadian position. Of course, in Whitehead's philosophy, this problem is handled by God: the past is preserved as objectively immortal in the consequent nature of God; moreover, the past affects the present because of the function of God in the arising of actual occasions. The final result of allowing problems of temporality to be solved by the apparatus of the eternal is perhaps brought home most forcefully by Edward Pols in his well-supported and well-argued claim that ultimately any so-called power of actual entities itself reduces to the power of eternal objects in certain kinds of relatedness to them. As he gradually unfolds this problem, it can be seen that it is implicitly linked to the notion of atomic, indivisible units of time.[42]

As indicated above, Whitehead was motivated by the need to avoid Bergson's seamless duration, the understanding of time as an absolute container, and the metaphysical bifurcations of the modern period, and to provide for the arrow of time in the face of the reversibility of the laws of science and the multiple time frames of relativity theory. But the alternative to Bergson need not be the move to ontological discretes. The alternative to time as an absolute container need not be the ongoing emergence of temporal atoms as the building blocks of time. The alternative to the metaphysical bifurcations of the modern period need not be the claim that the spatiotemporal point relations of science are obtained by extensive abstraction from the spatiotemporal relations founded in discrete units of duration. And the alternative to the reversibility of time need not involve the absolute fixity or determinateness of temporal atoms successively perishing to become an absolutely fixed, determinate past.

Pragmatism's own process metaphysics will be seen in part III to be structured by alternative solutions to these same problems that motivated Whitehead. But these differing solutions can be briefly highlighted here through a telling assumption on Whitehead's part that any satisfactory theory "must embody an understanding of the interweaving of change and permanence, each required by the other."[43] Yet the determinate and the temporal seem strangely at odds with one another. What is ultimately at issue is whether the fully permanent, the fully definite, the fully fixed has any place in a view that takes time seriously, or whether the notions of permanence, fixity, and definiteness devoid of all remnants of contingency and changing potentialities, of a discrete succession of temporal atoms in opposition to continuities of becoming, are due to ingrained habits of thinking that must be relinquished if temporal process is to have its due say. These crucial differences between Whiteheadian and pragmatic process will for the most part be put aside for now, however, in order to pursue the issue of temporal discreteness in some of its various manifestations.

Before turning to related issues in other process philosophers of the Whiteheadian lineage, it should perhaps here be stressed that the above analysis fully recognizes the accepted claim that Whitehead's position in *Process and Realty* is a prime example of the way in which no one kind of entity is fully explicable without reference to the other kind, and hence no basic concept is fully intelligible without all the other basic concepts, which must be grasped altogether in their interrelation before any one of them can be adequately understood. But the various problems connected with Whitehead's understanding of time, and the issues of discreteness and continuity with which these problems are intertwined, do not result from artificially abstracting one feature of Whitehead's position from the others. Rather it is in the need for their necessary interrelatedness that problems emerge, and coming to grips with these interrelations does in fact provide the underlying problematic for much of contemporary debate among Whiteheadians concerning the viability or possible reconstruction of his position.

Dewey expresses this problem from his own pragmatic perspective in his objection that "two methods seem to run through Whitehead's philosophy," one which makes distinctions within experience, and one which provides generalizations which are "morphological and static," which express "an aboriginal structure, the components of which are then deductively woven together." This latter gives rise to difficulties because of "the intermediary apparatus required in the interweaving of elements; the interweaving being required only because of the assumption of original independence."[44] In a different but related vein, the problems involved in obtaining creative advance by weaving together ontologically discrete atoms are reflected in Mead's observation that the ingression of eternal objects into events is an attempt to get contingency from spatiotemporal relationships that do not adequately allow for contingency and emergence.[45] In addition to what he views as the inadequacy of Whitehead's solution, Mead's objects in principle to his use of eternal objects as an illicit metaphysical hypostatization of an abstraction. Extensive abstraction "as a method of analysis and integration . . . asks for no other justification than its success. But Whitehead uses it as a method of metaphysical abstraction. . . . This seems to me to be an improper use of abstraction, since it leads to a metaphysical separation of what is abstracted from the concrete reality from which the abstraction is made, instead of leaving it as a tool in the intellectual control of that reality."[46]

A creative attempt to combine the pragmatic focus on continuity with the Whiteheadian focus on atomicity is to be found in Robert Neville's writings. His discussion is rich with perceptive insights concerning temporality in its various dimensions, and his distinction between conditional and essential features[47] as applied to past, present, and future provides a unification of time in terms of its conditional features that gets beyond some Whiteheadian problems and brings the issue to a deeper level. The following comments, however, will be limited to his own twist to the continuing Whiteheadian problem of bringing together an ultimately truncated temporality, a problematic that occurs in Neville's position because of the essential features of past, present, and future.

Neville holds that each temporal mode, past, present, and future, has its essential features that characterize its "timeliness." But no mode of time can be what it is without conditional features from the other modes. Temporality, as opposed to timeliness, is the harmony of all the modes in their essential and conditional features, a harmony that constitutes time's flow. While the temporality of each mode of time—past, present, and future—provides one form of conditional togetherness for the three kinds of temporality, this is inadequate to provide a togetherness of the different essential natures of past, present, and future.[48]

For example, the essential character of the past is its everlasting fixity, which does not touch the essential inner decisive dynamism of the present, nor the essential normativeness of the future.[49] Conversely, the essential dynamism of the present neglects the everlasting fixity of the past. The temporality of the present unifies the three temporalities as inherited past and future potentials for present action. This does justice to the essential dynamism of the present, to the conditional readiness of the past to influence the present, and to the conditional structures supplied by the future for possible actions. But it neglects the essential objective, everlasting fixity of the past. As Neville summarizes the problem, the modes are related to each other by the conditional features. But with respect to the conditional features, their essential features are mutually external.[50]

Hence he concludes that "we must indeed appeal to eternity to understand time. For there is a togetherness of the modes of time that is not a function of any of the kinds of temporality."[51] Rather, the essential features of time are together in eternity. If the truncation of temporal modes is in any sense allowed, then an appeal to eternity may well be needed to provide the togetherness. But as long as the assumption remains that the arrow of time requires an everlastingly fixed, finished, dead past, then the more one moves toward a dynamic, novel present, the firmer must become the truncation of temporal modes.

The generality of the issue as developed above in relation to time is partially indicated by Neville in his assertion that the inability of each of the three modes of temporality to unify them into one temporality is an illustration of an "important characteristic" of the general problem of the one and the many, and in the present context this translates into the problem of the relation of continuity and discreteness. This can perhaps be more forcefully illustrated by focusing on his view of time from the slightly different direction of his understanding of the flow of time as composed of temporal atoms. As he states, "There is a least unit of time in a things' temporal flow."[52] An atomic, indivisible unit of time for Neville is of course not pointlike, but rather is an indivisible extension or temporal dimension. Moreover, he stresses that the flow of time is smooth: "At no time is there a break between the finish of one emerging present and the beginning of the next; at the next time there is already a new emergent becoming. At no time is there a transitional bump from one present to the next."[53] In this way, "The flow of time is continuous."[54] This "arising" of continuity from the putting together of atomic units is encapsulated in his claim that the atomic moments of the present become continuous as they are integrated by the past. In this way, "the past of a person is not the set of individual decision points but the history of deeds and accumulated character of the individual in context."[55] However, from such a welding together of atomic units there can arise only a pseudocontinuity of contiguous parts

The Whiteheadian ultimacy of the ontological discrete is further evinced in his characterization of the two dimensions of the richness of the world at a level of metaphysical abstractness.[56] Things are infinitely "associated" ontologically and also infinitely deep or dense harmonies—infinite in their components units.[57] Just as time is ultimately composed of atomic building blocks, things in time are understood in terms of harmonies of infinite atomic components or units. Similarly, time is the conjunction of past, present, and future,[58] and though these are eternally together, they are together in the relationship of conjoined parts.[59] In his attempt to synthesize pragmatic and Whiteheadian insights concerning continuity and discreteness into a unified position, Neville *must* ultimately fall into one camp or the other, for the insights of the former ultimately negate the insights of the latter. The former offers the "emergence of" quasidiscretes within continuity, while the latter the "coming together of" ontological discretes to form a pseudocontinuity. One cannot have it both ways.

While Neville's philosophy is pervaded by a keen insight concerning the difficulty for the Whiteheadian perspective of bringing together disparate aspects of time, yet the need for "the-bringing-together-of" is never really questioned. The intuition of "the coming together of" discretes pervades and in large part defines Neville's philosophic stance.[60] Not only is this fundamental intuition and its ramifications that which leads to his move to an eternal unification, but it also structures the way the eternal provides a togetherness by the conjoining of the mutually external natures of temporal modes as well as the atomicities that are the ultimate components of the world's harmonies.[61]

The depth, pervasiveness, and illusiveness of the differing intuitions of Whiteheadian and pragmatic-style process philosophers can be found by turning to respective criticisms offered by each to the other. Not surprisingly, Charles Hartshorne, from his Whiteheadian lineage, finds Peirce's most serious mistake concerning time, and hence concerning his position in general, to lie in his emphasis on continuity.[62] As he summarizes, "Peirce's 'Synechistic' theory of experience as temporally continuous . . . made it impossible for him to conceive a definite unit of experience having definite predecessors and successors."[63] Hartshorne argues that "if discontinuity is not actual, there can be no unit events, but only arbitrarily designated slices of process, as short as one wants to take them. . . . What then about our awareness of a sequence of experiential presents each with its past and future?"[64] He attributes this difficulty here to Dewey as well, and holds that their positions suffer from the same problem that is to be found in Bergson's analysis.[65] Hartshorne holds that without ontological events as units of experience that are definite and atomic, there is no way to ground temporal succession.

Thus, according to Hartshorne, Peirce's position cannot account for "the asymmetry of time, time's famous arrow."[66] The principle surgery for this is a reconstruction that excises "that half of Synechism which pretends to legislate for actuality."[67] What is required is a metaphysical separation of continuity and discreteness, to be understood in terms of temporal modes. "Continuity is the order of what *can* happen, the future as such, not of what does happen. If so, actual becoming is discontinuous, and only the possibilities for becoming are continuous. . . . In some strange way, throughout his mature period, he missed seeing that if continuity is the order of possibility as such, its distinctive form, it cannot also be the order of actuality as well. How could he have failed to maintain this distinction?"[68]

He acknowledges Peirce's own understanding of the present as "infinitesimal" rather than as either finite or instantaneous temporally but finds this unintelligible. For Peirce would have to "hold that we go through an infinity of (unconscious) operations of remembering, and interpreting what we remember, even in the smallest stretch of time."[69] As he illustrates, "obviously the person reacts to his own past a good many times in the morning. How many times?"[70] This criticism, however, reasserts the very point of contention and misses the mark of the crucial difference between them. In Peirce's understanding of continuity and infinitesimals, such a "how many" of discrete instances is itself unintelligible, for as will be discussed in part III, Peirce holds that continuity is precisely the absence of ultimate parts in what is divisible. With ultimate parts one has at best only a "pseudocontinuity"[71] or a contiguous series.

The elusiveness of these radically different understandings can be brought into view by turning to the way in which Dewey's own approving appraisal of Whitehead misses the mark of the crucial difference between them, a difference that can be gleaned in Dewey's very statement of his approval. Dewey holds that the discrete-continuous relation may well have been the primary impetus in the development of the distinctive elements in Whitehead's position. But, "In any event, the need for reconstructing the doctrine of independent Newtonian atoms, which was forced by the doctrine of relativity as well as by that of quanta, brought the problem of atomicity-continuity into the foreground."[72] As he states the issue in his own terms, individuality is a "phase, though a decisive and outstanding one, of a process having continuity."[73] He reinforces this view with the direct quote from Whitehead: "It is a moment of sheer individuality, bounded on either side by essential relativity."[74] But the ontologically discrete atomic units of "sheer individuality" of which Whitehead speaks and the "phase of continuity" of which Dewey speaks are far from one and the same. The different "intuitions" that the above commentary both houses and hides is the difference between the "coming together of" ontological discretes to provide a quasicontinuity, and the "emergence within" continuity of quasidiscretes. It is the difference between an understanding of the discreteness of actuality as an ontological ultimate and as an idealized abstraction. From the backdrop of the issues highlighted thus far, the analysis will now turn to what is generally considered "the other side of the great divide" in ongoing metaphysical debate. That is, of course, substance philosophy, probably best represented on the contemporary scene by the position of Paul Weiss.

In developing his substance philosophy, Weiss distinguishes between two independently functioning, irreducible types of realities, Actualities and Finalities, the latter being irreducible permanent realities incorporated by actualities.[75] Actualities are the individuals in which Finalities are expressed. He lists five types of evidence for actualities as substances. First, appearances qua appearances are inert, yet they defy our epistemic imposition of boundaries by the pull on them "originating from beyond those appearances." Second, an appearance is subject to "coarsening," a density that, though continuous with the appearance, is distinct in nature and function. Third, certain appearances have a relationship of belonging together that they do not have with other appearances of the same type, leading to a common source upon which they converge. Fourth, though our actions are directed by appearances, they are "efficacious but incompletely so" because they reach insistent actualities that terminate and resist them. Finally, we are aware that we do not know all there is to be known; "beyond anything we know, there is always an unarticulated concreteness promising more to know. We come to know an actuality by symbolizing along the route of this unarticulated dimension."[76]

Yet none of these examples gives evidence of substance as opposed to process but rather of a "substantial" universe in the sense of a thick or dense universe, one that exhibits persistence, resistance, and efficacy. Indeed, Weiss himself, after stating that every actuality is a substance, goes on to speak in terms of substantiality,[77] and to continually stress characteristics that need not be at odds with process philosophy if process philosophy is not equated with the Whiteheadian variety. But Weiss recognizes no other type of process philosophy[78] and continually opposes his substance view only to the Whiteheadian alternative of an event ontology.[79] As Weiss continually characterizes his specific objections, these events or actual occasions do not occur through time but at a time. As he states, "some contemporaries have thought it desirable to replace substances with events. These, strung together, are supposed to yield a semblance of persistence, continuity, and distinctiveness—the supposed characteristics of substances. But such an account fails to yield anything that might act."[80] This is because "events just are"; no event can "persist and react." "Events occur, but it is substances that perform, move, change."[81] Weiss does not accept traditional notions of substance, stressing that the alternative to traditional views of substance need not be a theory of Whiteheadian events, but "another account of substance."[82] Substances, for Weiss, are persisting temporal objects that, as such, can handle the problems of endurance he finds in Whitehead's position.

While Weiss claims that substantiality requires actualities or substances, yet within his philosophy, it seems to be not substance or actuality but rather the finality he terms existence that allows the characteristics he desires. He holds that "each actuality exists apart from all the others. Each fills out a present of the same kind and duration, but in its own distinctive way, to make this its own present, having nothing to do with any other."[83] He clarifies the role of existence:

> Existence is what makes the Actuality effective, dynamic, able to act and interact. . . . By virtue of the continuity of the Existence in the Actualities with the Existence which lies between them, the Actualities are related to one another across an extended space. As the whole of Existence moves on, it implicates those Actualities; these, so long as they maintain a hold on their Existence, will move into the future concordantly. Despite the fact that every Actuality exists in itself, it is, by virtue of the continuity of its Existence with the rest of Existence, not only a being which keeps abreast in time with independent contemporaries, but which interplays with them, interacts with them to constitute a single slab of present cosmic time.[84]

Thus, it is the Existence that an actuality encloses that is the source of its vitality, "grounds space and time, and exhibits cosmic laws,"[85] allows actualities to be open to experience,[86] and provides the continuity that relates separate, distinct actualities to one another.[87]

The above quote indicates that what is involved in the relation between actuality and existence is the issue of the relation between discreteness and continuity and the question as to which is more fundamental. And, ultimately for Weiss, as for Whitehead, discreteness wins and the problematic becomes the welding together of discretes. In *Privacy*, Weiss speaks of an aboriginal, active continuum.[88] Yet "The irreducible units which condense it are to be understood in both Democritean and Leibnizian terms, the one emphasizing the bodily side of them, the other the private."[89] The entire discussion then proceeds in terms of indivisible atomic units, either atoms or atomic monads.

This issue is interwoven with Weiss's understanding of time. Weiss continually speaks of time as a continuum and recognizes as well that the flow of time cannot be a simple continuum, but rather requires "breaks." His way of handling this issue of time can be seen in his observation that "The impression remains that time is one continuous flow, though in fact it is a sequence of fixities displacing one another without interval."[90] An atomic unit or "fixity" for Weiss is not pointlike, but rather is itself an "indivisible extension."[91] "Time is a sequence of distinct, indivisible, extended units. An extended time without these units would be a simple continuum."[92] In this way, there is a "continuum of atomic instants."[93] Although Weiss explicitly denied this, yet ultimately within his position discreteness is primary, be it the units of time or the "units" of substance, resulting in a quasicontinuum of contiguous discretes, in discrete elements that must come together, even though "in a sequence without interval."

This attempt to understand the continuity of time in terms of indivisible atomic units leads to a radical truncation of past, present, and future in Weiss's position. Weiss denies that there is a direction to past time in and of itself, but only in terms of an external future.[94] This future that introduces directionality must be external to the past, for the absolute fixity of the past requires that it have no directionality; as past, it is immune to temporal flow. Further, in relegating possibilities solely to the future, Weiss holds that while they are "directive of things in the present," yet "These possibilities are external to the things in the present." Possibilities provide an "external future" for persisting actualities.[95] Without determinations by this external future, there would be only a plurality of present ongoings indeterminately related to one another.[96]

In being external to actualities or temporal substances, there is no "logical space" for possibilities which fail to be realized in Weiss's system. Rather, possibilities are indeterminate (the possibility of "either A or B"), and are made determinate as incorporated in actualities. (Either

A occurs or B occurs.) An actuality makes concrete and determinate the more inclusive, less determinate "inevitably realized possibility."[97] The import of this point is indicated by the fact that the crucial problem of providing for possibility as external to the present is what first led Weiss toward finalities.[98]

In light of this brief focus on Weiss's position, it can be seen that though his substance philosophy and Whiteheadian process philosophy are usually seen as diametrically opposed, and in fact Weiss's arguments for the substance perspective frequently unfold as an opposition to Whiteheadian process, yet there are remarkable similarities between these two positions. First, both hold that possibilities or potentialities have an ultimate status apart from actuality in such a way that they must remain in some sense independent of what transpires in the present. Hence Whitehead has eternal objects or pure potentialities that are unaffected by the limitations placed on "real potentiality" by the ongoing succession of perishing actual occasions; Weiss has possibilities that are inevitably realized regardless of what transpires in actuality, though actuality makes them progressively less inclusive and more determinate. And for both this ultimate fixity of possibilities is intertwined with the need for the absolute fixity of the past in order to preserve the arrow of time.

The broad entrenchment of this type of thinking about possibility can be seen in the way it permeates Irwin Lieb's development of a view of time as a rejection of Whitehead and Weiss alike.[99] Because Lieb never questions the validity of the identification of possibility with the need for a "mighty reality"[100] in some way eternal, fixed, and noncontingent, but yet rejects this latter, his own temporalist philosophy rids itself of the reality of possibility. Thus according to Lieb, "there are no real possibilities."[101] Rather, "individuals and time are the basic realities." Because he rejects rather than reconstructs the notion of possibility, he again must utilize the absolute fixity, definiteness, and actuality of the past[102] to preserve the arrow of time.

Second, in spite of Weiss's objection that Whiteheadian atomic events are actualities that just "are" and cannot themselves act or interact, and hence the necessity of substance, yet his own substances partake of his criticism. For what makes actualities effective, able to act and interact, is not their nature as substances but the continuity of Existence that relates them. Finally, underlying all of the above is the crucial similarity that explains the others: While both positions are concerned with continuity, yet in the final analysis both ultimately incorporate a continuity that has the function of bringing together or welding together or binding together ultimately discrete units of actuality in an ultimately discrete temporal process, producing a pseudocontinuity of discretes. In this way, both positions stand together on the opposite side of pragmatic process, separated from the latter by the great divide of different "intuitions" of time and the role of discreteness and continuity in its unfolding.

In anticipation of following chapters, one further implication of the explorations of this chapter should be noted, as its significance will take on added weight as this work progresses. Whitehead, Weiss, and the various members of their lineage are all concerned with the embeddedness of humans and their experience within a rich nature and the need to overcome the destructive subject-object split that has plagued epistemology. Thus, for example, focusing on the kinship between human experience in general and nature, Nicholas Rescher can characterize the Whiteheadian position as one in which "human experience constitutes the model or ideal type of the processes that characterize natural reality in general. For Whitehead saw the core nature of reality as typified by the living experience of an 'actual occasion'" in its full immediacy.[103]

Yet as discussed earlier, when it comes to the issue of time, Whitehead holds the view that the continuity of the stream of experience is a surface feature prominent for consciousness, but the underlying reality is a succession of atomic experiences, actual entities or actual occasions of experience. Hence, our lived sense of time, time consciousness, is in some way subjective, set over against an alien objective reality of successive temporal atoms. The lived sense of time and the objective reality of time are incompatible, and the gulf between them is ultimately unbridgable.

Whitehead, Weiss, and Neville alike each present not only a grand view of the indeterminate richness of nature and the reintegration of humans and nature, but also a clear recognition of the importance of cosmic time in understanding any such integration. Yet for each the human experience of time is uniquely alienated from the "natural reality" of time.[104] The temporal continuity of the flow of time consciousness is the most fundamental and pervasive feature of human experience, yet its nature as seemingly continuous closes us off from the cosmic reality of time as constituted by discrete temporal units. Indeed, this fundamental alienation of human awareness and cosmic reality in large measure accounts for the fact that while the basic intuition of cosmic time and related issues for Whitehead, Weiss, and Neville alike is the "bringing together of" discretes, their respective understandings of human experience and knowledge are pervaded by the fundamental holistic intuition of "emergence within."

As will be seen, pragmatism shares with all these positions a fundamental focus on the indeterminate richness of nature, the holistic understanding of experience and knowledge, the reintegration of humans and nature, and the crucial role of cosmic time in any such undertaking. But by maintaining the temporal continuity of lived experience and the fundamental intuition of "emergence within" in its move to cosmic time, pragmatism moves this endeavor beyond the impasses structured by the building blocks of the discrete.

More immediately, however, the following chapter will explore the way discreteness and the problems it engenders pervade diverse attempts to rid philosophy of its concern with a metaphysics of time.

CHAPTER 2

Derrida, McTaggart, and the Demise of Time

This chapter will examine two positions that involve the rejection of the possibility of a philosophical concern with a metaphysics of time in order to elicit certain similar relevant assumptions operative in the disparate arguments. The focus will turn first to Derrida's deconstructive position growing out of the continental tradition, and then to McTaggart's argument, which, emmeshed as it is within the analytic tradition, is today usually debated in terms of the logical analysis of temporal propositions but which ultimately concerns the underlying metaphysical issue of the reality of time.

At first glance, bringing Derrida into the discussion of discreteness and continuity in understanding time may seem a bit misplaced, since Derrida clearly proclaims that time as continuous and time as discrete are equally metaphysical concepts that emerge from the play of *différance* and should be discarded. Yet an examination of Derrida's position reveals that one cannot escape the significance of time, for it is his own implicit stance concerning the nature of time that gives rise to his deconstructive position and the central role of trace in its structure. The role of trace comes to the fore both in understanding the living present and language. And it is a particular understanding of the role of trace in the living present that undergirds an understanding of the role of trace in language, for traces are constitutive of signs in the same way that the protentions and retentions as traces are constitutive of the living present. In clarifying this relation, however, it will prove beneficial for the sake of clarity to proceed from trace in signs to trace in the living present, which governs the former.

Derrida recognizes as an essential dimension of signs their semiotic function as standing in for, as taking the place of.[1] Further, signs separate human experience from the experience of pure presence as pure "unmediated immediacy." For Derrida, this indicates the downfall of presence. This downfall of presence has two dimensions. First, there is no "presence to consciousness," in its constitution of time; second, there is no "presence to things," no semantic reference of language as tied to the world.

The basis of Derrida's separation of signs and presence begins to emerge through his modification of Saussure. Rejecting Saussure's radical distinction between signifier as the material or physical image and signified as thought or concept expressed, he holds that the meaning of the sign emerges from its difference from other signs, and the signified can itself become a signifier. The meaning of a sign is not immediately present as Husserl and Saussure thought, but rather is constituted by a "tissue of differences," a network of referrals, and every so-called simple term is marked by the trace of another term.[2] Meaning transpires in the "play that is the web of language,"[3] a play that weaves a network of traces. Since everything transpires within language constituted by the network or system of signifiers, and is locked within that system, there is no escape from the system of signifiers and traces.

Thus Derrida finds the signifier to be related to other signifiers in relations of difference rather than to be dependent on a relation to any particular signified, and holds the system of signifiers to be inescapable. In so doing he asserts a new and radical sense of syntax that is far more than merely syntax as form in contrast to content.[4] This has been referred to as a "syntax of syntax" from which the "formal syntactic properties can be syntactically composed and decomposed."[5]

This reduction to syntax liberates the signifier from the "oppressive regime"[6] of presence as immediacy, and at once ties it to the time flux over against structure or meaning. For if syntax is prior to semantics and there is a "syntax of syntax,"[7] then the flux of syntax, the diachronic, is in no way tied to or subordinated to semantics or to meaning, or to the structure of the system of language.[8] This movement away from structuralism in favor of the flux underlying meaning and language involves the centrality of time in relation to signs. The discussion will now turn to this issue.

Some of the most scrutinized passages by deconstructionists in their recent attention to Husserl's analyses of time are the passages where the distinction between the living present and the instant or the punctual now is made. What will be important to bring to light by turning to Derrida's focus on Husserl is not his criticism of Husserl per se, but the implicit understanding of time from which it arises and the way in which Derrida's implicit point of reference structures his own deconstructionist position, for his approach here provides the basis from which he pulls the rug out from under not only the phenomenology of Husserl but, indeed, all philosophy.

Husserl's account of time addresses the Humean problem of a flux of nows discrete from one another and that, as unidimensional, allows for no depth. Against this backdrop, Husserl specifically focuses upon the duration of an object of awareness as the same from moment to moment. His analysis reveals the protentions or anticipations, and retentions or fundedness, that constitute a network of intentionalities and the living present as a duration. In this way the living present has depth and continuity, making the presence of the object possible.

Husserl's account of the living present entails a thickness to the "now" grasped in a quasi-immediacy that does not require, and must be radically distinguished from, a separate act of remembering, which would attempt to repeat and reproduce the occurrence, a secondary memory. Rather, the retention and consequent endurance by receding can be considered a kind of primary memory, a presence that is not the result of repetition or reproduction. Yet in his analysis Husserl stresses the beginning point of something appearing in time consciousness in a way that lends itself to the interpretation of the "now" as an instant point. Although it is explicitly clear that the now in Husserl's account cannot be collapsed into the pointlike instant, the ambiguity of his treatments gives rise to a tension between the living present and the instant, the "thick now" and the "point now."[9]

Derrida, following his deconstructive method, exploits this latent tension within Husserl's account between the "instant as a point"[10] and the thickness and depth of the present. For Derrida wants to wedge a separation in the duration of the "thick now," making two alien and discrete parts. According to Derrida, the spread of the now is for Husserl described on the basis of the self-identity of the now as point, as a "source-point."[11] From this perspective, Derrida questions and deconstructs the thick now or living present, realizing that with a point "now" the presence of the perceived present can appear as such only by containing a nonpresence. He goes on to argue that perception constitutes the pure point now for Husserl and "memory every other phase of the continuity" of the living present.[12] All memory, then, is considered nonpresence similar to the nonpresence of secondary memory. This relation to nonpresence according to Derrida "radically destroys any possibility of a simple self-identity. And this holds in depth for the constituting flux itself."[13]

That nonpresence and otherness are internal to presence strikes at the very root of his deconstruction of Husserl's living present. As Derrida states, "The difference between retention and reproduction, between primary and secondary memory, is no longer the difference, intended by Husserl, between perception and nonperception; it is rather a difference between two modifications of nonperception."[14] Hence, for Derrida, Husserl's "thick now" of the living present collapses in the face of the alterity of past and future, revealing the failure of Husserl's time constitution. For Derrida, the presence of the present is the result of the "bending back of a return,"[15] of repetition, of différance, and not the reverse.[16] Thus Derrida holds the position that "this pure difference, which constitutes the self-presence of the living present, introduces into self-presence from the beginning all the impurity putatively excluded from it. The living present springs forth out of its nonidentity with itself and from the possibility of a retentional trace."[17] In fact, the relation to nonpresence actually makes possible the presence of the perceived present.[18] Even the point instant gains its identity from its relation to nonpresence or nonperception.

The above analysis shows that Derrida picks up on the wrong side of the tension between the living present and the instant in Husserl's ambiguous treatment. Hence, for Derrida, Husserl's "thick now" collapses in the face of the alterity of past and future, revealing the failure of Husserl's time constitution. The flux, at Derrida's hands, becomes not a continuity but a "pseudocontinuity" of discretes. The trace functions in the nonidentity of the present with itself, a present including a complete alterity because a thick, continuous living present inclusive of past and future has been transformed by Derrida into a living present composed of a unification of discretes, a present and a not present; at the very heart of the living present is complete alterity, and within this context the trace functions. While time as truly continuous gives rise to a living present or "thick now" that can be characterized in terms of neither absolute identity nor absolute difference, Derrida offers the pseudochoice between absolute identity or absolute alterity: either a present that is the pure presence of sheer immediacy, or a living present, a "thick now," already necessarily constituted by discrete elements having the relation of complete, dire alterity or otherness. Derrida rightly rejects absolute identity, leaving him with the remaining alternative of absolute alterity. His very setting of the alternatives for understanding the living present, or the thick now, demands and presupposes a view of time as ultimately discrete, a "pseudocontinuity" of discretes.

Hence, according to Derrida, rather than excluding "impurity" as supposed, the self-presence of the present, in being constituted by pure difference, is structured through its nonidentity with itself, and is possible only through its relation to nonpresence. In this way, it is not just potentially a trace, but "is always already a trace."[19] The living present, constituted by an identity with itself and the sketched-out-ahead of the future and the retained sketch of the immediate past, provides what Derrida refers to as the intimate relation with its "outside,"[20] as an openness to exteriority. And it is differance that "makes the movement of signification possible only if each element that is said to be 'present,' appearing on the stage of presence, is related to something other than itself but retains the mark of a past element and already lets itself be hollowed out by the mark of its relation to a further element."[21] The living present takes on the role of "power of synthesis" and of the "incessant gathering up of traces."[22]

The trace, then, relates to the future and to the past as protention and as retention, and constitutes the present by its relation "to what it is not,"[23] that is, the trace constitutes the present by its relation to what it is not—as a future or past—and, as such, not even as a modified present. In fact, the relation to nonpresence actually "makes possible"[24] the presence of the perceived present; furthermore, even the instant gains its identity from its relation to nonpresence or nonperception. This nonpresence is the place of the trace, for protention and retention, which are for Derrida both on the side of nonpresence, are traces that come to constitute the living present.

The living present, then, is constituted as a nonsimple and derived effect, as synthesis of trace, retentions and projections. The use of language involves the future and the past in the interplay of traces of differences within the play of différance to bring about the present and its presence. This play of traces is a "sort of inscription prior to writing, a protowriting without a present origin."[25] For Derrida, the trace is in some illusive sense the origin "of sense in general," it is the "différance which opens appearance."[26]

The way in which Derrida's understanding of language shares the consequences of his fragmented view of time is well encapsulated in his claim that "signification is formed only within the hollow of differance: of discontinuity and of discreteness, of the diversion and the reserve of what does not appear. . . . This hinge [*grisure*] of language as writing, this discontinuity, could have, at a given moment within linguistics, run up against a rather precious *continuist* prejudice."[27] Différance, then, makes signification possible because of the interval that separates the present from alterity. In this way nothing precedes différance.[28]

Derrida's dismantling of the self also shares the consequences of this fragmented view of time. As he states, the subject, "depends upon the system of differences and the movement of différance"; "the subject is not present, nor above all present to itself before différance"; and "the subject is constituted only in being divided from itself . . . in temporalizing, in deferral."[29] Thus he can claim that "the self of the living present is always a trace,"[30] and proclaim "the total absence of the subject."[31]

Derrida, in breaking up temporal passage into a play of discrete moments, forces a choice between signs without presence or presence without signs. Recognizing the importance of signs, he accepts signs without presence. Hence there should be little wonder that Derrida loses both speaking subject and semantic reference.

But precisely the now as an instant point makes his view of trace unworkable. For if the instant now is discrete and isolated, then the trace is separate from it. If the claim is made that the system as a whole at any moment is constituted by the interrelation of all the elements, it must be argued that all the elements at this instant are the constituted effect or product of protentions and retentions, of traces, which are not present, so that the enigma stands of the instant being vacuous as a point, and discretely cut off from any other instant. Protentions and retentions, since they are constituted effects, cannot bring about the required unity, but rather presuppose it. How is the trace to cross over from one empty now point to another? It is not connected to any, except insofar as it is in the rich present state as a trace of a past with no present. Can such crossing over of the gaps or chasms between instants be allowed by the presupposition of primordial discreteness that underlies this deconstructive view of the system of signs and trace? Since the trace is itself an effect, it cannot bring about a unity of instants except as itself already an effected synthesis. Yet such a unity must preexist trace for its very possibility, for trace involves more than the instant. This difficulty in Derrida's view of trace points toward the importance of a view of time, and hence of signs founded upon it, that rids itself of all vestiges of discreteness.

Derrida's understanding of the functioning of signs, operating as it must within the framework of the above pseudo-alternatives, is incapable of allowing for presence as mediated by meaning or for a temporal span, in which presence is rooted, that can be understood neither as dire alterity nor as pure identity. He rightly rejects pure identity but, in accepting the remaining option of dire alterity rooted in discreteness, accepts an alternative that renders impossible the very constitution of the most basic effects of the play of différance, effects that must provide the tools for the linkage of discretes: trace and the protentions and retentions it involves.

Hence in spite of the central role of trace in Derrida's deconstructive stance, he can provide no adequate basis for it. Discreteness of the instant or an atomism of time cannot support any connection between two times or two instants of time. In Derrida's position there cannot be established any such connection between two now points any more than in Hume's position there could be established a real connection or causal relation between two now points discretely cut off from one another. Derrida's deconstructionist stance, implicitly founded in a temporal discreteness, is the destructive but inevitable final result of the philosophical assumption of time as discrete. But, as shown earlier, that same assumption renders inadequate the tools his deconstructive stance needs to maintain itself. Oddly, then, Derrida's own deconstructive exploitation of vestiges of time as discrete, evinced even by those who are often most anxious to avoid it, points out both the impossibility of the effective functioning of trace within his own deconstructivist stance and the importance of the viability of such a concept in some form, whatever the label given, for any adequate view of time. While Derrida's deconstructive stance explicitly rejects time as a metaphysical concept, then, yet his position is structured throughout by an implicit acceptance of time as discrete. From this backdrop, the focus will now turn to McTaggart's quite different and seemingly unrelated attack on a metaphysics of time to elicit some key similarities.

McTaggart's argument for the unreality of time has been the ongoing source of debates, usually within the linguistic confines of tensed versus tenseless propositions. Yet there are underlying metaphysical issues that are brought out more clearly in McTaggart's early essay than in the writings of later participants in the ongoing debate. The explicit focus of McTaggart's position concerns the relation of the experience of time to cosmic time. Is the experience of temporal passage an experience of a "real" property of the cosmos external to the experiencer or does it belong "merely" to the realm of the experiencer? Intertwined with this explicit focus, however, is again an implicit but pervasive assumption concerning the nature of time that lends support to his arguments.

McTaggart analyses time into two series of events, which he calls the A-series and the B-series. The A-series is the succession of events passing from future to present and then to past, while the B-series is the same succession as housing the unchanging relations of earlier, simultaneous, and later. The events in both series are the same, we never observe events in time except as forming both these series. But while in the A-series the temporal description changes, in the B-series it does not. Pointing out that it would be universally admitted that time involves change, McTaggart then goes on to argue that change without an A-series is not possible, for there is only one class of characteristic of an event that can change, the class consisting of the determinations of the event as past, present, or future, the terms of the A-series. Whatever other qualities an event has, it always has. But past, present, and future constitute incompatible determinations, yet every event must have them all. To try to avoid the contradiction by an appeal to present, past, and future moments that occur at different times merely transfers the problem of events to that of moments. And there can be no B-series, which involves the time relations of earlier and later, without the A-series, since without the A-series there is no time. Thus, as he concludes:

> The reality of the A series, then, leads to a contradiction, and must be rejected. Moreover, since we have seen that change and time require the A series, the reality of change and time must be rejected. And so must the reality of the B series, since that requires time. Nothing is really present, past, or future. Nothing is really earlier or later than

> anything else or temporally simultaneous with it. Nothing really changes. And nothing is really in time. Whenever we perceive anything in time—which is the only way in which, in our present experience, we do perceive things—we are perceiving it more or less as it really is not.[32]

This assumption that an event is fully determinate and hence cannot change, thus requiring the exactly same thing to have incompatible predicates, incorporates the notion of time as discrete. An event for McTaggart is a discrete, in this case a "discrete substance," which can be compounded to form a compound substance as a "larger" event, but which cannot itself change.[33] Past, present, and future, then, become mutually exclusive attributes of atomic, fully fixed units. Hence for Mctaggart as for Derrida, the claim for the demise of time is parasitic upon of the assumptions of discreteness and fixity.

McTaggart's position has given rise to an ongoing war between A-theorists and B-theorists in the attempt to hold onto the reality of time or, perhaps more precisely put in the context of their ongoing debate, to determine whether tensed sentences can be translated into tenseless sentences or, in other related terms, whether tensed or tenseless temporal characteristics describe time as it "really" is. A-theorists argue that our experience of time includes more than the experience of the B-relations of earlier, simultaneous and later. It includes the experience of the mind-independent A-properties of past, present, and future, with the focus on the immediacy of the Now and the sense that it moves through time. Only by the movement of the instantaneous Now, which "shifts inexorably, independently of what we do"[34] can passage be explained. From this direction, the focus of debate becomes whether we do or do not experience a mind-independent property of presentness that provides the sense of the now point. The A-theory is put on the defensive here because it cannot specify what this property of presentness is over and above the features allowed by the B-theory, such as simultaneous with, before, and after.

The argument for the A-series is made from a different direction by holding that McTaggart's inconsistency concerning the A-theory arises from his particular and misplaced conception of passage.[35] We should conceive passage as succession, as "the modal sequence of *first* future, *then* present, *then* past."[36] This first/then succession uncovers the nature of our A-theoretic impression and points out the inadequacy of the B-series. But this type of position is put on the defensive because the Now, which is temporally unextended, occupying only a single point in time, must also be thought to extend over many distinct instants. Accordingly this argument is accused of in fact showing that the inconsistency of the A-series is not merely some idiosyncrasy of McTaggart's conception of passage, but rather is an inconsistency for any A-theoretic conception of the moving Now.[37]

From the perspective of Robert Brumbaugh's arguments, the problems for the A-theorists can be placed in a slightly different context that highlights certain other aspects of the conceptual problems involved in the subjective sense of time. While agreeing that there is no distinguishing feature for the present, he holds that the problem lies in the very notion of the present, for the present or the Now, as the "locus of transformation for two realms—future and past—" leads to the "problem of understanding how an existent thing lies simultaneously in its future and past."[38] Since Brumbaugh is "certain that the law of identity must hold literally for all reality,"[39] he turns to the notion of a substantial identity that persists through changing aspects that are not constitutive of its reality, a concept that is made intelligible through an understanding of time as a "field within which such things as paths of motion can be graphically conceived and understood by single equations or formulae."[40] This position clearly asserts the ultimacy of absolute identity and atomic substantiality that, in McTaggert's position, characterizes the event.

B-theorists defend the reality of time by turning McTaggart's order of priorities around and arguing that the B-series provides for real time, which is in fact tenseless time, while the A-series, as subjective, is in some way derivative from or caused by or reducible to, the real tenseless time.[41] B-theorists hold in common the distinction between real time and subjective time, thus cutting off our sense of time from the reality of time, our experience of tense from the reality of tenselessness. The strength of the position stems from the weakness of the A-theory in its inability, first, to give an account of the passing "Now" that allows for the reflective awareness of the immediacy of a present devoid of past and future and that claims its independence from our experience of it; second, to overcome the conceptual problems of in some sense extending an unextended temporal Now over a series of temporal points, which is necessary for a truly moving Now.

What both sides of the issue hold in common is the assumption that the Now, if it be at all, must be a moment of pure immediacy, a pure present devoid of past and future immediately "there" for or within our reflective or cognitive grasp, or the assumption of the absolute identity and discreteness of what is ontologically ultimate, be this developed in terms of the Now, the event, or changeless substance. The A-theorists seem to recognize the precritical, prereflective, unanalyzed experience of temporal passage, but have attempted to understand it within the categories of discreteness and self-identity, as well as the related categories of immediacy, pure presence, and the real as the cognitively apprehended, as a result distorting the unanalyzed sense of temporal passage beyond recognition. The B-theorists, drawing on the paradoxes with which the A-theorists are faced when the sense of time is dismembered by the above theoretical categories, accept the separation of our subjective sense of temporal passage from "real" tenseless time that can accommodate the atomistic succession of unextended time points; thus they either explain this subjective sense away in terms of "real" tenseless time, or release it to lead an unexplained subjective life of its own, in this way utilizing and reinforcing the self-defeating but well-entrenched distinction between subjective experience and objective reality.

The problem with which A-theorists are faced in their search for the apprehension of the presentness of the present in their attempt to hold on to the reality of tensed time is not limited to the framework within which the above ongoing debate takes place. As an example, Errol Harris, in a book rich with insights concerning temporality in particular areas, searches in vain for the awareness that distinguishes the presentness of the present, what is "now" devoid of past and future, and saves the coherence of temporality only through an atemporal ordering. And this search for the mark of the present that distinguishes it from past and future is intertwined with his understanding of the role of discreteness and continuity as it relates to temporality. For time is a "consecutive series of events," and as such requires a nontemporal generation and can be held together only by a "totality which is not a member of the series."[42] Change is real, but its reality is possible only because of the continuity of the eternal, which both generates and holds together the members of the discrete series. This eternal provides not only the ontological but epistemic glue for temporality, for the temporal series can be cognized only as a whole, and as a whole it does not change; as a whole it is grasped in a single apprehension by a unitary, cognizing subject.[43]

The ultimate reality of temporal change lies in the succession of atomic events within an eternal, changeless continuity, and the ultimate apprehension of temporal change is a timeless apprehension of a series of events—an eternal presence. Ultimately, then, the relation of discreteness and continuity implicates a time that makes sense only if "all time and all process imply, and exist only within, a nontemporal totality . . . time is always an image, a progressive manifestation, of eternity.[44] In his discussion of Zeno's paradoxes, Harris has an insightful discussion of the needed interrelation of discreteness and continuity in dealing with the issue of temporality. As he concludes, "The appropriate answer to Zeno is that continuity and discreteness are inseparable and interdependent moments of every quantity."[45] Yet the structure Harris offers is that of temporality as discrete succession, with the "continuity" reduced to a totality or collection of discretes and itself immune to time.

While Harris's metaphysics of eternity is far removed from the problematics of A-theorists and B-theorists, his position reflects a remarkable parallel of problems and solutions that characterize the tensions between the two sides of the analytic debate. Like A-theorists, he searches for the mark of the present and winds up in paradoxes as he dismembers our sense of temporal flow with theoretical categories of static cognition. And, like B-theorists, he draws on these paradoxes to introduce a separation of our subjective sense of temporal passage from "real" tenseless time that can accommodate the atomistic succession of unextended time points. For time as image of eternity seems, in its own way, to undermine the very reality of time as experienced, creating a chasm between our lived sense of time and the ultimate nature of tenseless reality. What these parallels seem to reinforce is the claim, already made to some extent and rounded out in the brief comparative overview that concludes this chapter, that seemingly radically diverse contemporary perspectives have shared conceptual roots deeply embedded in the soil of discreteness and fixity, and these perspectives develop through their diverse ways of coming to grips with the family-related sets of dilemmas and paradoxes that arise as these conceptual roots run up against the sense of temporal flow embedded in ordinary human experience.

In bringing together the various threads within the above discussion, Derrida and McTaggart are more similar than they may at first appear from the standpoint of their alien traditions. While Derrida can unravel time and with it philosophy in general because his underlying presupposition of discreteness leads to a radical alterity, McTaggart can condemn time to the "unreal" because his underlying assumption of discreteness leads to a radical identity. But these are really opposite sides of the same coin, for discreteness and the alternatives of absolute identity or absolute alterity go hand in hand. It seems again, this time from a different direction, that an adequate understanding of time must rid itself of the notion of the fully discrete and, along with it, the notions of absolute identity and absolute alterity.

Further, language for Derrida and the sense of temporal passage for McTaggart are equally groundless, for each of these is cut off from any anchor in reality or from any reference beyond itself. This becomes logically necessary because within their respective frameworks any grasp of reality requires the presence of the fully determinate and discrete, be it fully structured and determinate things or facts incapable of accommodating the functioning of signs or fully determinate atomic events incapable of accommodating different temporal modes. This atomicity of events, with the immediacy and pure presentness of the Now it invokes, pervades the attempts, within the tradition of McTaggart, of A-theorists and B-theorists alike to assert the reality of time in some sense, as well as some of the attempts, as exemplified by Harris, to save the reality of time by a move to the eternal.

Before leaving this chapter, some themes that relate it to the previous one, the problematics of which were set by the Whiteheadian perspective, should perhaps be clarified. What must be emphasized is that the problems of time to be found in the positions of Derrida and McTaggert cannot be juxtaposed with those of Whitehead and others previously discussed for several major reasons.[46] First, Whitehead is concerned with the nature of cosmic time, while Derrida refuses to allow us to step out of language much less speak of speculative cosmic issues, and confines his understanding of time to its constitution in consciousness; and McTaggart can lead us only to the alleged unreality of time. The entire problematic is radically different. Second, Whitehead's whole stress is on the durational, nonpointlike nature of the present, while the pointlike instant-present gives rise to the radical alterity running through Derrida's position as well as the paradoxes asserted in McTaggart's. Third, Whitehead has utilized the philosophical tradition to construct one of the grandest metaphysical schemes to emerge in this ongoing tradition, and its force will forever exert a positive impact on, and give direction to, the future of the tradition, regardless of problems one may find with its particular doctrines.

Meanwhile, McTaggart himself offers only a philosophical dead end, around which later A-theorists and B-theorists attempt to maneuver. And Derrida has utilized the tradition to show that it destroys itself. Some may think that it needs to be destroyed; some may think that out of the ashes a better road for philosophy can be built. But, while this latter view can make use of some of the numerous keen insights operative within Derrida's position, what he himself offers us is not reconstruction but deconstruction. As David Wood makes this point, Derrida's position "is successful in exposing, in its own terms, the metaphysical motifs in (philosophical) texts. What it cannot do . . . is to offer anything like another positive account of anything, let alone the subject of our concern—time. But the fact that deconstruction cannot do it does not mean it cannot be done in a way responsive to its insights."[47] As Richard Kearney queries concerning the plight of philosophy after deconstruction, "Where do we go from here? How may we hope to ever escape the endless self-parodying of postmodernism which announces the 'end' of everything but itself?"[48]

The plight in which deconstruction has left us is well encapsulated by Richard Bernstein's query at the conclusion of his highly sympathetic reading of Derrida, "What are we to do after we realize that all archai tremble? It is almost as if Derrida wants to bring us out the wilderness, wants to uberwinden the violent 'history of the West' without providing us with an orientation for avoiding the abyss of nihilism that he so desperately wants to avoid." This highlights "a problematic which is—if we are intellectually forthright—not only Derrida's, but *our* collective problematic."[49] Though Bernstein earlier noted that "Derrida himself deconstructs (takes apart) this stark contrast: fixed center or chaos and formlessness,"[50] Bernstein thus also acknowledges that in ridding philosophy of fixed centers Derrida has provided no constructive path for avoiding the nihilism of chaos and formlessness.

All this having been said, however, there is, in very broad strokes to be sure, a striking similarity in the respective dilemmas of Whitehead and Derrida in trying to provide a continuous link between ultimately discrete temporal units, a dilemma addressed by Hume's skepticism and perhaps unresolvable in the face of the ontological ultimacy of a temporal atomism of any kind. Moreover, while Whitehead might seem to progress from a prereflective sense of temporally founded human experience to the temporal features of reality in general, as discussed earlier he in fact begins with the model of atomic particles, and from temporal atoms attempts to explain the human experience of temporality. Because of this, though Whitehead's philosophy throws us out onto a temporal universe, the continuity of the stream of experience as a "surface feature prominent for consciousness" is itself a type of "alienation" from the "real," for the ultimate reality of time is that of a succession of atomic actual entities. This disparity between the consciousness of time and real time, as well as the atomism of time that underlies it, provides the kind of thinking that is exploited by McTaggart in his argument for the unreality of time, as well as the alternative attempts to hold on to the reality of time by separating real time and subjective time. In Derrida's hands, temporal atomicity causes language itself to lose its link to any cosmic moorings, for the alternatives of pure identity or pure alterity, giving rise to the alternatives of either presence without signs or signs without presence, results not only in there being no temporal presence to consciousness in its constitution of time, but also no "presence to things," no openness onto the world, no semantic reference of language as tied to the world.

What is lost by all through their respective understandings of time as discrete is the ultimate import of the sense of time embedded in the prereflective, "felt" textures of the praxis orientation of concrete human existence.[51] The next part will turn, in various ways, to Heidegger's focus on the sense of time embedded in concrete human existence and some implications which lie implicit within it. Following that, part III will focus on the pragmatic understanding of the nature and ultimate import of the sense of time embedded in our ongoing praxis, exploring the way in which this understanding avoids, through its own *re*constructive path, both fixity and chaos.

PART II

The Heideggerian Alternative and Its Pragmatic Connection

CHAPTER 3

Heidegger and Temporal Existence: The Pragmatic Contours of an Expanded Context

Each of the previous positions was seen to begin with temporal atomicity and in so doing drives a wedge, intended or not, between the sense of concrete human existence and the building blocks from which it supposedly arises. Heidegger dwells on the import of the sense of primordial time embedded in concrete human existence, but in so doing fails to adequately recognize its significance for understanding a broader sense of temporality, though the direction for doing so lies implicit in his position. In what follows it will be argued that Heidegger's position lends itself, in fact can be interpreted as implicitly progressing toward, a rethinking of the temporal dynamics of a cosmic process in which humans are at home.

The manner in which this chapter draws out these implicit tendencies will take its shape from the direction of classical American pragmatism as it will be developed in part III. For now, what is important to note is that classical American pragmatism lies in radical opposition to the type of Anglo-American pragmatism that Mark Okrent brings to his understanding of Heidegger, and hence it is not surprising that the position of this chapter as well as the remaining chapters of this part will run counter to the general thrust of his interpretation of Heidegger's link with pragmatism.[1] For example, Okrent finds the key to Heidegger's pragmatism to lie at least in part in the fact that Heidegger cannot root his view of temporality in his analysis of the authentic being of the human but instead must rely on a transcendental semantics. As a result, according to Okrent's understanding of pragmatism, Heidegger's semantic-pragmatic bent led him to suppress the latter part of *Being and Time*, and the "turning" in the later Heidegger's thought is in fact an attempt to escape the pragmatic elements of his thinking. Thus, Rorty's explicit disagreement with Okrent regarding the reading of part of *Being and Time* is an in-house one that revolves in part around "Okrent's statement that all pragmatism must either be based on a *transcendental* semantics or be self-contradictory."[2] Given this Anglo-American or neopragmatic interpretation of American pragmatism, Daniel Dahlstrom rightly views pragmatism as a possible source of the recent neglect of Heidegger's discussion of temporality.[3]

For classical pragmatism, however, concrete human existence contains a crucial inroad for understanding temporality not only in its existential but also in its cosmic significance, and what will be developed shortly is that the "turning" of the later Heidegger brings him even closer to the pragmatic position by embedding human existence in a broader context. From this perspective, and with the exclusion of the controversies and interpretive directions surrounding the pulling of Heidegger's thought toward a "pragmatism" quite alien and in fact largely contradictory of classical American pragmatism, the following discussion can begin to unfold an interpretation of the temporal direction of Heidegger's thought.

Heidegger views the history of philosophy as metaphysics to be the history of the interpretation of the privileged evidential role of presence, intertwined with an interpretation of the privileged temporal role of the "now" of the present. For, he holds, since the pre-Socratics, Being has always been viewed, in some variation of ways, as presence. By the time this temporal determination of Being came to be reflected upon by Aristotle, the understanding of time had been taken over by physics.[4] Time, as the horizon for interpreting the meaning of being, disappeared from view. As Heidegger encapsulates the problem, "Since all metaphysics from the beginning of Western thought, Being means being present. Being, if it is to be thought in the highest instance, must be thought as pure presence, that is, as the presence that persists, the abiding present, the steadily standing 'now.'"[5] The metaphysics of presence as yielding immediate evidence is understood by reference to a determinate temporal mode, the present now in which it occurs.

By contrast, Heidegger seeks to develop an existential view of time as distinct from the metaphysics of presence and the privileging of the present. Only through a recovery of original time or temporality as the primordial ontological structure of human existence and the horizon for the inquiry into Being can there be a recovery of what was lost. This leads him to the development of a view of time that characterizes the temporal unfolding of concrete human existence, but in so doing he in his own way cuts this off from the cosmic reality in which it is embedded. Because of this he is frequently criticized for the existential limits he imposes on temporality. However, what will here be asserted is that Heidegger's position lends itself, in fact progresses toward, a rethinking of the temporal dynamics of a cosmic process.

Heidegger holds that everyday living in the world loses itself among the objects of its concern and hence fails to see the horizonal structure that makes these objects of concern possible. Our ordinary concept of time is that of a series of sequential now-points, an indefinite series of "nows," where the future consists in the nows that have not yet come but will be, the past consists of the nows that once were but are no longer, and the present is the now that at the moment is. In this common view, however, the structure and significance of world-time,[6] its origination in primordial time or temporality, and temporality itself have all been "covered up." The true dimensionality of time, as a "reaching throughout" has been lost. But even this everydayness views time as irreversible, and this is the result of primordial time that underlies and filters through the everyday understanding of time as a sequence of now points, an understanding that has covered up primordial time. Everydayness provides both the source of temporal significance and the concealment of it. The ordinary concept of time is the result of a covering over and a leveling down of primordial time. Heidegger proposes to uncover this primordial time in his existential understanding of time.

For Heidegger the unity of temporality comes into view as the structure of the human as such. Each dimension of temporality—past, present, and future—is a reaching throughout, and in this way "time constitutes the metaphysical continuity of Dasein."[7] Each dimension has its proper horizon, and the unity of the directions a unified horizon. The reciprocal unity of these three-dimensional horizons in temporality is at once the structural unity of care as the human mode of being in the world. The human exists in the world as care, and the ultimate meaning of care is temporality. The examination of the human as care discloses a threefold unity of its structure that is founded in its constitution as temporality. "Temporality makes possible the unity of existence, facticity, and falling, and in this way constitutes primordially the totality of the structure of care," for each has its respective representation in future, present, and past.[8]

For Heidegger, the very structure of the human way of being founds or projects world as a significant whole or as the horizon for meaningful existence. The human mode of being *is* being-in-the-world. Conversely, world exists when, and only when, humans exist.[9] The human way of being as temporally constituted thus enters into the temporal character of world. Indeed, the unified horizon of all temporality provides the condition of possibility of world as a holistic meaningful context.[10] The human being-in-the-world is the "clearing" where things can be disclosed and appear as meaningful. The everyday world is grounded in the structural process of the human way of being as being-in-the-world, and within this everyday world the unities of things that emerge into presence lie in their functionality, not in some underlying essence.[11]

The horizon or meaningful context for a certain kind of experience must be projected or developed in advance, must be opened up, before that kind of experience can take place. Anything emerges in our world, enters into our awareness, only by an anticipatory preparation that opens an horizon of meaningfulness. Anything emerges in our world only as projected upon or within the context of such a horizon. What we already understand within our world can appear to us and be understood by us only because we have already established a more fundamental encompassing frame of reference, field of interest, or holistic context. Anticipation is ingredient in the fundamental nature of all understanding, and any interpretive explanation is the development of the possibilities "there" because of a more fundamental context or frame of reference. This more basic frame of reference or horizon opens a realm within which possibilities emerge for consideration as possibilities to be utilized. Even the most primordial grasp is not a presuppositionless apprehending of something that reveals itself, for the meaning of things is constituted, receives its significant structure from, our fore-having, fore-sight, and fore-conception.[12]

Further, the nature of world as a functionally significant whole or a meaningful totality at once throws light on being.[13] Beings, "entities," things, are encountered in the world only as their emergence within human experience is made possible by the meaningful totality that is world. Even the projections or meaning bestowing activities implicit and concrete in our everyday concernful engagement with things in the world involve a pre-hematic, nonobjective awareness of what it is to be. And, if what it is to be is understood, this requires a horizon or context for that understanding.

For Heidegger, the sense of being, which must be distinguished from particular beings or entities that reveal themselves within the world, can only be discerned by first discerning the sense of being as the being of the human. The understanding of being is to be found only in the constitution of the human. As mentioned above, the human is essentially an anticipatory drive in terms of its potentialities. It is not some substantive entity in time, but rather is temporally constituted as its own potentiality, and all understanding is rooted in temporality. The being of the human as temporal founds the frame of reference for the understanding of being. Hence the basic context for the understanding of being is time. Being is understandable only within the basic context or horizon of temporality. Temporality, then, is the source of the structure of the being of the human, of the sense of being that belongs to the human, of the world as a significant whole or unified context, and of the emergence of things as functional unities within the world.[14]

As discussed earlier, world as a context of significance is essentially founded in the human mode of being. In the early Heidegger, being itself is described as the "total meaningfulness of the world" projected by the human. Being "is dependent upon the understanding of Being."[15] The later Heidegger himself attempts to correct the implications of his numerous statements in *Being and Time* that seem to conflate being with the human "lighting" of world. As he queries, "But does not *Being and Time* say on p. 212 . . . 'Only so long as Dasein is, is there Being'? To be sure. It means that only so long as the lighting of Being comes to pass does Being convey itself to man. But the fact that the . . . lighting as the truth of Being itself, comes to pass is the dispensation of Being itself. . . . This sentence does not say that Being is the product of man."[16]

The question of the famous turning in Heidegger's thought, and the debate as to whether or not it constitutes a rupture or a continuity between the earlier and later Heidegger, is by now a well-ploughed field. The following discussion both presupposes and illuminates the view that there is no rupture, but rather a continuity in which each directional stance helps explain and clarify the other.[17] The early Heidegger is concerned with being as the "total meaningfulness of the world," and the coming to pass of being as the luminosity of the world in Being and Time occurs through human "care." The later Heidegger replaces the human as the locus for the unconcealedness of Being, but just what direction this takes is open to debate. Some have argued that in the later Heidegger language replaces humans as the place for the unconcealedness of Being. Language becomes the so-called "house of being" and humans are its shepherd. Humans live in language, which provides the advent of Being itself.

Yet the later Heidegger can be understood to replace the focus on humans with a focus on the broader being process as the cosmic context within which humans are emmeshed. The ensuing discussion will pursue the far-ranging implications of this. In the later Heidegger, the nature of humans as world founding becomes explicitly understood in terms of the broader context of being. He continues to develop an understanding of time more basic than that of a succession of "nows," but turns to its development in a way that is to a large degree independent of the existential focus on the human way of being in the world. Heidegger, however, rejects the notion of a suprasensible reality that grounds appearances as in substantive realism or in any variation of the noumenal-phenomenal break, or that houses them as in modern subjectivistic epistemology. The broader context of being is that within which humans are embedded as participants. Here the focus on being, as that which gives possibilities, is not on the being of humans operating within a meaningful frame of reference but on that which grounds the human mode of being in a broader context. The concreteness of the possibilities engendered by this extantness or independence of being[18] can never be exhausted by human disclosive activity. It makes possible both disclosedness and its "other side," opacity. In brief, it is a "thick" independence that is not exhausted by our apprehension of it.

The human horizons of interpretation must be grounded in independent being, which has its own potentialities for being disclosed to us as exhibiting certain types of structures. Heidegger, then, can emphasize a certain essential "powerlessness" of the human mode of being.[19] Yet this independent reality cannot be understood in terms of a collection of independent entities having essences that we just come to know, for the significance of things in the world is not independent of the meaningful context established by human activity. The "natures" or "essences" of things in the world reflect the character of the human and the character of what is there independently of humans. What is extant or independently there is not *a* being, but is that rich matrix of being within which individual things can emerge as functional unities within the context of world as a holistic horizon of significance.

World, as seen earlier, is not a collection of intraworldly things, but rather the context of meaning within which worldly things can reveal themselves in their significance. Similarly, the independent matrix of being, as having its own potentialities for exhibiting certain types of structures, is not a collection of individual things having essential natures; rather it is that matrix which discloses itself through the emergence of individual things as it enters the worldly horizons of significance or the worldly horizons of human purposive activity. Things as they emerge within the context of world are a function of the matrix of being which has its own developing potentialities and possibilities, and the potentialities and possibilities projected through the structure of human activity. Only through temporally structured activity does the matrix of being let itself stand out as individual things, and likewise, in our encounter with individual things we are experiencing the being of these things. The interactive unity of these two poles that constitutes temporally grounded experience points toward the temporality of each.

There is growing debate among Heideggerian scholars as to whether or not primordial or originality temporality is reducible to the authentic temporality of care.[20] While the conflation of authentic temporality and original temporality still dominates the interpretive scene,[21] there is growing recognition that this conflation may not hold. This latter position is encapsulated in the claim that temporality has "some content above and beyond care" and that "the three-fold constitution of care stems from the three-fold constitution of temporality."[22] In this way, our study of care "introduces us to a matter whose scope outreaches care."[23] Yet the view that primordial temporality cannot ultimately be grounded in care does not lead scholars to see this as a way to move to a cosmic temporality that is there independently of the human, but rather they interpret this as a critical problem that leads Heidegger away from phenomenology toward some transcendental condition as the supplier of primordial temporality.

This tendency is found in claims as diverse as that of Okrent, discussed above, that Heidegger moves to a transcendental semantics, and the claim that such a separation would require a Kantian transcendental. As this latter has been expressed, what would be this content that is above and beyond care "if not some sort of purely formal entity or condition such as Kant's 'pure intuition'? But this would imply that Heidegger has left phenomenology behind and is now engaging in establishing a transcendental framework."[24] This refusal to follow a path toward a sense of that which is there independently of the human may be due to the general tendency of this tradition to at once recognize that cosmic time as scientific time cannot be turned into a metaphysics of time, yet to allow for no other sense of cosmic time. This other sense of cosmic time, such as that which will be offered by the pragmatic approach, is one that is captured by neither scientific nor metaphysical hypostatizations of abstractions but rather by a move from the phenomenological features of the temporal richness of concrete human existence toward a focus on the temporal richness of the cosmic dimension that these features embody.

That which is there independently of the human or human activity, and the kind of temporality it involves, can best be brought into focus by turning to Heidegger's discussion of that-which-regions.[25] That-which-regions provides the open-ended availability of possibilities. It grounds all that is and all the ways of being, and is itself present in all that appears within the world. It grounds the human mode of being, and provides the possibilities that allow the functional unity of things to emerge in relation to human creative, purposive activity.[26]

That-which-regions stresses the openness and activity of being. Heidegger develops the view that the openness of the region[27] that is the ground of the human's opening must, like the human, be understood as active. To bring this active feature into focus he calls it at times, "regioning."[28] That-which-regions, regioning, is the dynamic ground which brings forth the essential dimensions of the human. As Heidegger stresses, "The question concerning man's nature is not a question about man."[29] As a result, in addition to the three dimensions of time already mentioned, Heidegger later introduces a fourth dimension that makes possible the other three in the structure of human existence. "But the dimension which we call the fourth in our count is, in the nature of the matter, the first, that is, the giving that determines all."[30]

The interactive unity of both poles within the structure of human experience can be clearly seen in Heidegger's discussion of the relation between that-which-regions and horizon. What grounds the human way of being, what lets the horizon be, is not something other than the horizon, but rather horizon emerges as this ground enters into the disclosive activity of the human. The human emerges within that-which-regions as that manner of being in reference to which that-which-regions is disclosed via the mediation of world as the horizon of meaningful encounter with it.[31] In encountering it, we partially "mold" it to our human demands, letting it stand out as isolatable objects for our use.

Things that emerge in our worldly experience are in part a function of "the dynamic unfolding of that-which-regions" and in part a function of the way we get hold of it, for its indeterminate possibilities are structurable in alternative ways, though it also constrains these ways, accommodating some but not others. Our world, then, opens in one direction toward an independent dynamic matrix of being and the possibilities it presents, and in the other direction toward the structures of our modes of grasping it and the possibilities such modes of grasping allow. Thus, the temporal features of world are intertwined not just with the temporal features of the human mode of being but with the temporal features of the ongoing cosmic activity of "that-which-regions." The reaching throughout of temporality, and hence the pervasiveness of possibilities throughout, provides the indeterminate nature of cosmic reality that we render precise within our world through our modes of anticipatory activity.

This dynamic reality both reveals and conceals. As the side facing us, as horizon, it reveals, but even as it reveals it conceals, for its ontological "thickness" is not exhausted in its horizonal relation to the human. As Heidegger states, "The horizon is still something else besides a horizon. Yet after what has been said this something else is the other side of itself, and so the same as itself."[32] It can be named, and hence thought about[33] but it cannot be represented "insofar as representing everything has become an object that stands opposite us within an horizon."[34] It cannot be described, for "any description would reify it."[35] Yet Heidegger does describe it, but he describes it as that which is not itself an object, which cannot be reified into an object, but which, as the ground for the emergence of the human and hence for the emergence of the possibility of its own worldly significance, has certain characteristics that flow into, are consonant with, the dynamic, temporal structure of humans and the world in which they dwell. The whole dynamic interplay, the whole transactional context, of this finite temporal unfolding, which involves the reciprocal interplay of appropriation,[36] is being. Heidegger, then, can hold that it can be said neither that time is the product of the human nor that the human is the product of time.[37] It is this dynamic interplay that "calls for" thinking.[38]

Before turning to the nature of this "thinking," our intertwining with this dynamic matrix will be further explored by turning to a concept that at first glance seems alien to and even contradictory of Heidegger's problematic, that is, naturalism. Indeed, the explicit focus on the context of naturalism within pragmatic philosophy has been a key factor in the historical alienation of pragmatic philosophy and the thought of Heidegger. For Heidegger explicitly undercuts the "natural attitude," with its construction of an ultimate reality of scientific contents and its scientific approaches to the nature of the human, in order to make manifest the fundamental existential structures that found them. Thus, for example, Heidegger attacks what he considers "the blindness and arbitrariness" of pragmatism's scientifically grounded position.[39]

Yet part III will show that the pragmatic rethinking of temporality leads to three senses of nature. First, there is nature as that "thick," dense, independently real that is the foundation for all that is and for all the ways of being, including the uniquely human having of a world as the horizon of meaningful activity. Second, there is nature as our worldly environment, as it emerges within the contours of our world, as it emerges as a network of relations of things used in everyday purposive activity, as it emerges in the meaningful contours by which we cut into the temporally structured processes of nature. Third, there is nature as an object of science, which we theoretically cast upon our everyday natural environment as a second-level reflection upon it.

Neither the richness of nature as that from which all, including humans, spring, nor the richness of nature as the human everyday worldly environment, can be confounded with or reduced to the abstract character of the events and objects in the world of science. Nature in its indefinite richness is the foundation for our everyday worldly natural environment; nature as a system of scientific events and objects is a reflective abstraction from it. Further, the first sense of nature is not something "other than" the latter two, but rather the latter two are diverse ways in which this nature enters the openness of human horizons of activity. Hence it founds the latter two. Moreover, only because independent nature enters into human purposive activity as the everyday worldly environment can nature as the object of the abstractive focus of the scientific endeavor begin to emerge. This threefold understanding of nature will provide a perspective from which to view Heidegger's various statements.

Heidegger almost always speaks of nature in the derived, abstract sense of that which is grasped through the focus of science. He is concerned to show that the content of this focus, upon which previous ontologies of nature are based, loses the characteristic features of being-in-the-world grounded in the ontological structure of world as a significant whole, as a meaningful totality of possible involvements rooted in human anticipatory activity. These foundational dimensions make possible the very being of nature as an object of science. The natural attitude, which attempts to give a metaphysical ultimacy to the derived objectivities of science, is oblivious to the fundamental ontological level that grounds them. Yet Heidegger at times refers to nature not in this derived sense, but as that which is discovered in the meaningful network of relations that constitutes the environment as our "closest world." He observes that when nature is defined by science, then "the Nature which 'stirs and strives,' which assails us and enthralls us as landscape, remains hidden."[40]

Again, he points out that nature as intraworldly, as that which is discovered within the world, can be thought of as scientifically uncovered in a theoretical, physicochemical way or as the "'nature out there,' hill, woods, meadow, brook, the field of wheat, the call of the birds."[41] In this latter sense, nature reveals itself within the totality of our involvements with what is ready-to-hand, with what serves as "equipment" within the context of everyday purposive activity. This meaningful network of everyday involvements is what has been stripped off in the narrow view of nature as grasped by the theoretical gaze.[42] Thus Heidegger characterizes the nature that comprises the object of physics as the "unworlded world,"[43] for it has lost the character of the world in which we dwell.

There emerges even in the early Heidegger a third sense of nature. Being-in-the-world belongs to the being of the human as human. However, intraworldliness, as what is within the world, belongs to the being of nature "*not* as a determination of its being but as a *possible* determination . . . one that is necessary for the possibility of the uncoverability of nature."[44] Nature is something "extant," that is, has its being independent of the human,[45] though intraworldliness "devolves upon" it when humans exist.[46] As Heidegger succinctly summarizes, "World is only, if, and as long as a Dasein exists. Nature can also be when no Dasein exists."[47] This nature is not "other than" the nature of our everyday worldly environment, but is that which becomes our worldly environment as it enters the domain of the human: "World-entry and its occurrence is solely the presupposition for extant things announcing themselves in their not requiring world-entry regarding their own being."[48] Nor is it "other than" the narrow nature of the scientist; rather, it is that which can enter the world of the scientist through a second level theoretical focus founded upon nature's discoveredness in the everyday environment.

Here the objection may arise that though nature has its independent being for Heidegger, this is not relevant to his fundamental ontology as an uncovering of the existential structure of the human or as the disclosedness of being. Because Heidegger so strongly reacts against the recurring understanding of an independent nature in terms of the reification of scientific contents, he never refers to nature in any attempt to characterize the broader being process. The independence of nature in this broadest sense, however, as foundational of all that is and all the ways of being, is applicable to being as well as beings, and is ultimately inextricably linked with his fundamental ontology as an uncovering of the existential structure of the human or as the disclosedness of being that can never be exhausted by human disclosive activity. This can be brought into focus by turning briefly to his concept of earth, which indicates that the embeddedness of humans in nature is as essential to their being and to the understanding of being as is their being-in-the-world. This assertion requires that we again move from the earlier to the later Heidegger and see the extantness of nature in its relation to the above features of that-which-regions: the dynamic temporal ground that brings forth the essential dimensions of the human, and that, as structured through our temporal worldly horizons, manifests itself as the things we encounter.

It has been seen that for Heidegger, to refer to that which is independent of human disclosive activity as nature would be to employ language in a way that furthered the tendency of past philosophy to illicitly ontologize the contents of science as "ultimate reality." Yet in indicating the interplay between world and that "independently there" from which all springs, including the human way of being as being-in-the-world, Heidegger makes brief but revealing use of the concept of earth. As Heidegger stresses, earth is not to be understood as a mass of matter, or as the astronomical idea of a planet.[49] Rather, his characterizations of earth give to this concept the function of that-which-regions. Earth is essentially the "sheltering agent" for all that arises.[50] The world grounds itself on the earth, and earth juts through the world. The earth cannot dispense with the open region of the world if it itself is to appear, to gain world-entry, while the world in turn "cannot soar out of the earth's sight if, as the governing breadth and path of all essential destiny, it is to ground itself on a resolute foundation," for the earth is that "on which and in which man bases his dwelling."[51] The earth conceals, but concealment "is not simply and only the limit of knowledge in any given circumstance, but the beginning of the lighting of what is lighted."[52] Thus, like "that-which-regions," earth represents, as "the side facing us," the horizon of the openness of world, but even as it reveals it conceals, for it is not exhausted in its horizonal relation to humans.

Heidegger characterizes the relation between world and earth as one of strife. Such an essential strife, however, is not that of disorder or destruction, but rather is a relationship in which the "opponents" allow each other to assert themselves in their essential dimensions, yet to assert themselves as belonging intimately to each other.[53] Though Heidegger usually refers to the strife between world and earth, he once alludes to their "counterplay."[54] This characterization seems more appropriate, and perhaps a better word would be "interplay," for there is a reciprocal interrelation between the two. Earth roots world, and world lights up earth.

Further, earth is as essential to the nature of the human as is world. To be human is to have a world; being human means being in the world. Yet being human means, also, "to be on the earth as a mortal. It means to dwell."[55] The earth is that on which and in which the dwelling of humans is grounded.[56] To dwell on earth does not mean to use it for technological purposes, nor does it mean to recognize, in reflective or detached manner, one's dependence upon it. Rather, to dwell on earth means to experience the natural embeddedness or earthiness of existence as essential to one's being. Just as the essential feature of the human as world founding has been concealed, is something essential that has fallen into oblivion, so in losing the import of dwelling on earth something essential has fallen into oblivion, "something decisive is concealed," for the earthiness of dwelling "is not experienced as man's being; dwelling is never thought of as the basic character of human being."[57]

Our embeddedness in nature is as essential to being human as is the development of a world as the horizon of meaningful encounter with nature. Heidegger demands a return to the pre-scientific horizon of lived experience, and his implicit contemporary naturalism makes this experience what it is. The primacy of our natural embeddedness is that from which arises both the distinctively human activity of projecting a world as the horizon of meaningful encounter, and the need for any world to accommodate the features of this natural embeddedness, be it the world of everyday purposive activity or the derived abstractions of the world of science.

Nature, then, as the foundation of all that is, of all the ways of being, is the dynamic temporal ground, the "regioning," that brings forth the essential dimensions of the human and, as structured through our worldly horizons, manifests itself as the things we encounter. As such, it is the ground of the dynamic interplay that calls for thinking. Heidegger refers to such foundational thinking as releasement to that-which-regions. This seems to be not so much a way of thinking as a way of experiencing. At this point the so-called mystical Heidegger may be thought to come into play. However, as Heidegger himself clarifies his so-called mysticism:

> But isn't all this unfounded mysticism or even bad mythology, in any case a ruinous irrationalism, the denial of ratio? . . . As long as ratio and the rational still remain questionable in what is their own, talk about irrationalism is unfounded. . . . The effect proves the correctness of the technological scientific rationalization. But is the manifest character of what is exhausted by what *is* demonstrable? Doesn't the insistence on what is demonstrable block the way to what is?[58]

Heidegger, in connecting the above with his reflections on Aristotle's claim that "it is uneducated not to have an eye for when it is necessary to look for a proof and when this is not necessary," is pondering "in what way that which needs no proof in order to become accessible to thinking is to be experienced."[59]

This accessibility lies in the attunement of "foundational thinking"; in experience attuned to the source of all that is, of all the ways of being; in the experience of the appropriative or interactive context of humans and the dynamic, processive universe in which they are embedded and which is part and parcel of their being. The attempt to bring such a process to language is founded on "attunement," and only on the basis of attunement "does language gain its tuning."[60] Only through the experience of attunement as the source of its own inspiration can philosophy regain its vitality, can it recover the experience that gives rise to the questions. And only in the context of this attunement can there emerge the full significance of Heidegger's asssertion that:

> Here we must in all sobriety understand clearly that temporality is in no way something that is to be beheld in some superabundant and enigmatic intuition; it discloses itself only in conceptual labor of a specific sort. But also it is not merely hypothetically supposed at the beginning without our having some vision of it itself.[61]

The focus on the later Heidegger leads to the view of some that he has returned to a metaphysics of presence he set out to deny. But the metaphysics of presence, of direct evidence, is tied to the time determination of the present now, and in working toward a different understanding of time, one that gets beneath the notion of time as a series of now-points, Heidegger works to a different understanding of presence. Primordial temporality provides the temporal features for an ontological "presencing" that is not static or inert but is dynamic and evolving. Presencing in its most general sense expresses the ongoing interactive temporal dynamics of humans with the universe in which they are embedded. Presencing allows for neither a presence yielding immediate evidence nor a present occurring in a tenseless now. There is no unmediated presence and there is no tenseless now. Presencing, as temporally dynamic, includes past, present, and future; and, as including past, present, and future it essentially requires the mediating function of interpretation to become the experiencing of meaningful things. To assert an intrusion of a metaphysics of presence into Heidegger's thinking, à la Derrida, is to ignore Heidegger's rethinking of time and, intertwined with this, his reconstructed understanding of the notions that cluster around it. For presence no longer excludes the mediating function of meaning, and the present no longer excludes past and future as an immediate "now." As John Caputto makes an analogous point from a slightly different direction, Heidegger's notion of nearness "has nothing to do with the metaphysics of presence, pace Derrida."[62]

The above way of focusing on Heidegger from the direction of pragmatic themes to be developed in the next part has attempted to show that he can be understood as implicitly progressing toward a rethinking of the temporal dynamics of a cosmic process in which humans are at home. For the sense of the temporal structure of the human mode of being is intertwined with the temporal features of the being of the extant or the independently real, yielding phenomenological features of experience as a function of the intertwining of each dimension and hence as manifesting the temporal features of both. While hints of this are given in Heideggerian thought, this direction is not clearly pursued, and probably for reasons deeply embedded in his thinking.

There is a deep ambivalence in Heidegger's writings. On the one hand, there is a sporadic, often implicit concern with the sense of nature not as any scientifically described universe, or as the world that is there through human time constitution, but rather as that natural order within which the human emerges; a concern with nature as that which we never exhaustively encompass in either our perceptions or theories, which is independent of the various worlds of perception and scientific theory but which nonetheless does not transcend them in the sense of being something that does not appear in experience.[63] Yet Heidegger quickly backs away from such a move for, as frequently happens, when he comes close to focusing on such a dimension, he seems to find in this focus a danger of absolutizing the contents of science, confounding his own halting tendency to move toward a cosmic temporality with the illicit reification of the eventful world of science.

This is intertwined with another subtle problem. For ironically, though Heidegger focuses on the lived body and its temporal structure, he truncates this from the idea of the body as an animal body tied to nature. As this is perceptively highlighted by Caputto: "Heidegger is clearly uneasy abut this animal body of Dasein's. It is hard to place." As Caputto details, Heidegger does not want human being or the human body thought in the framework of living being or life, but rather in the framework of providing a clearing or opening for Being's truth. He sees this as being a constant in Heidegger's thought ever since the Marburgh period when he replaced the expression "factical life" with "Existence" and "Dasein." Heidegger thus "read [the] 'life' out of Dasein." Dasein is not basically a form of life but a place for the emergence of truth.[64]

Again, it would seem that Heidegger's focus on the temporality of the lived body falls short of a move to integrate this with the temporality of a cosmic process because the body as an animal organism is ultimately understood in terms of scientific contents. His attacks on biologism or the reduction of the biological to scientific contents led him not to rethink the biological organism and the cosmos in which it is embedded, but to sever the distinctively human from its biological roots and the cosmos with which these roots are organically tied. As Caputto succinctly puts a parallel point, "Heidegger's treatment of the animal shows his hand. It verges on a Cartesianism which treats animals as little more than machines."[65] Heidegger's ambivalence concerning nature and the biological organism permeates, in analogous fashion, the issue of cosmic time. Thus, while Heidegger has tendencies that lead him toward an integration of existential and cosmic time, he has countertendencies that prevent a clear move in this direction.

One further point concerning the Heideggerian view of temporality should be noted that, from the pragmatic stance, becomes highly problematic. The counterpart of the temporal structure of care was seen to be found in Heidegger's unfolding of the temporal structure of understanding and disclosedness in general, and Heidegger gives a dominant temporal mode to each of the types of human disclosedness. However, this may be counterproductive to his overall understanding of primordial temporality as a "reaching throughout," for while this does not lead to a break between past, present, and future, it does introduce a truncation of function that is perhaps potentially too isolatable. This pragmatic objection is anticipated in David Wood's discussion of Heidegger's inadvertently having turned a condition of possibility into a discrete phenomenon. For as Wood queries concerning the implicit but illicit move toward discrete phenomena, "Could this not be claimed for his general account of disclosedness, and perhaps even of his descriptions of the temporal ekstases?"[66] Moreover, Heidegger's position reinforces the delimitation of potentiality and possibility in terms of temporal modes in a way that must be rejected by the pragmatic rethinking of temporality. The final move from time as discrete to temporality as a "reaching throughout," as well the final move to a thoroughgoing integration of lived temporality and cosmic process, requires a radical reconstruction that demolishes many of our ingrained habits of thought, including some of those still operative in Heidegger's thinking. Prior to turning to this endeavor, however, it will be useful to explore some further directions in Heidegger's philosophy, rooted in his understanding of temporality, which mark affinities with pragmatic philosophy.

But before proceeding further with this focus on Heidegger and pragmatism, a far-reaching possible objection to the present chapter as well as the two that follow must be addressed. One can argue that even if the interests of pragmatism and Heidegger may be interpreted to converge in the ways developed throughout this part, yet the historical contexts from which they operate are so diverse as to render any real comparison impossible. This work does not intend to deny the immense differences between their respective historical contexts. Nor, however, does it intend to embed the issue of temporality or the related issues of process naturalism, scientific method, and the a priori within these differences. Rather, the deliberate intent here is to abstract out from diverse historical contexts in order to focus on the remarkable convergence of the two traditions toward a common novel understanding of such issues. This convergence, which is deeply embedded within and pervasive of their respective philosophies—ultimately because rooted in converging understandings of time—has remained hidden precisely because of the radical differences between the respective historical contexts in which each philosophical position operates.

The further objection may arise that the direction in which the chapters of this part take Heidegger's position is set by the framework of classical pragmatism. This is indeed the case, but it is here held that while this interpretation may be selective of certain implicit tendencies to be found in Heidegger's thought to the exclusion of others, it is not distortive of Heidegger's thinking. Moreover, all interpretation involves selectivity of emphasis. The justification for the present path is that it casts a light on Heidegger's thought in a way that can make his insights concerning the temporality of concrete human existence open to a broader context of temporality such as that to be offered by pragmatic philosophy, thus making his philosophy a more fully rounded one. This direction, it would seem, offers a more fruitful path than that provided by present interpretive trends that find Heidegger retreating from phenomenology in a move toward transcendental semantics or a Kantian transcendental as the ultimate supplier of primordial temporality.[67] With the above clarification of the intent and method of this entire part of the book in mind, the ensuing chapter will turn to the understanding of scientific method in the positions of Heidegger and pragmatism and the temporal issues with which this is intertwined.

CHAPTER 4

Heidegger, Scientific Method, and Temporality: The Pragmatic Pathway for Deepened Dynamics

The claim was made earlier that the key to the integration of human and cosmic time lies in a refusal to reify scientific time as cosmic time or to accept any variation of ultimately discrete temporal units. If this is to be the case for Heidegger, then his understanding of the scientific endeavor must allow for this situation. This chapter will explore Heidegger's understanding of the scientific endeavor and the bond it establishes with pragmatic philosophy, indicating again from the direction of classical American pragmatism the way he stops short of a full utilization of his insights in this area.

The linkage of pragmatism with experimental method or the method of science as the model for understanding the nature of knowledge has been a key factor in the historical tendency to interpret pragmatism as a type of naturalistic reductionism that gives an empirical or natural or causal account of cognition based on the findings of science, and that reduces the human to lower levels of existence. Given this script, pragmatism again becomes the natural enemy of phenomenology, for phenomenology emphatically reacts against these reductionistic tendencies in order to focus on an irreducible field of meanings by which humans constitute their world and on the irreducibility of the qualitative fullness of primary or everyday experience. According to Heidegger, the dominance of reductivistic metaphysics based on the findings of science is directly linked with the objectionable features of pragmatism.[1]

The tendency to conflate the pragmatic focus on scientific method with one or another of the various scientific reductionisms rejected by phenomenology is probably nurtured by the general fact that the method of gaining knowledge that was the backbone of the emergence of modern science was confounded with the results of the first "lasting" modern scientific view—the Newtonian mechanistic universe. An illicit reification of scientific content, resulting from an inadequate understanding of scientific method, led to a worldview that gave rise to a quantitatively characterized natural universe and to either dualistic causal accounts of knowledge in terms of correlations between mental contents and material objects or reductionist accounts in terms of stimulus response, with the complete rejection of mentalistic terms. Heidegger's rejection of the "natural attitude" with its illicit reifications of the objects of science is in fact a rejection of all remnants of modern worldview frameworks and their evolving variations.

However, implicit in Heidegger's criticism of pragmatism is that same conflation of scientific method and scientific content that has plagued the philosophical tradition in one way or another since the rise of modern science, but that pragmatism emphatically rejects. The radical rejection by pragmatism of this conflation, a rejection unrecognized by Heidegger, allows the pragmatic focus on science as method to provide the foundation for its own criticism, similar to that of Heidegger's, of the modern worldview understanding of science and its various related frameworks. Pragmatism is indeed naturalism in that humans are within nature, but nature cannot be understood as the mechanistic universe of the Newtonian worldview. Rather, our primordial experience opens onto the prescientific richness of nature, which contains the qualitative fullness revealed in ordinary experience. Hence, pragmatic naturalism rejects the "natural attitude" to which Heidegger objects. What will be shown below is that their respective ways of rejecting the scientific reductionism of the "natural attitude" leads ultimately to, and in turn emerges from, common understandings of scientific method rooted in a shared vision of temporality.

While a misunderstanding of the nature of the linkage of pragmatism with scientific method has been a crucial factor in the historical philosophical alienation of pragmatism and the philosophy of Heidegger, the clarification of this linkage can accomplish several goals. First, it can help illuminate the way shared fundamental intuitions concerning temporality and continuity have implications for wide-ranging affinities between the two positions. Second, it can also help illuminate and enrich the significance of their respective attempts to understand, and to establish the proper relation between, abstract thought and concrete experience, between the abstract time of science and the primordial temporal flow that grounds it. It will be seen that the pragmatic understanding of scientific method[2] as the model for understanding the dynamics of experience requires a return to the temporal foundations of the everyday world in concrete human existence, the basic features of which are similar to those characterized by Heidegger, while Heidegger's phenomenological account of these temporal foundations incorporates his own understanding, similar to that of pragmatism, of the dynamics of scientific method. Each of these positions is rooted in a recognition not only of the temporal origins of our ways of understanding reality but also of the temporal continuity and indeterminacy that this involves. Third, this clarification can show the way Heidegger's understanding of the nature and function of scientific method naturally tends to deepen itself, via a pragmatic pathway, downward toward a phenomenology of temporal existence in general and from there toward a cosmic temporality of continuity and indeterminacy.

The ensuing examination will first turn to the similar dynamics of scientific method as found in pragmatism and in Heidegger's philosophy and then show, again from the pragmatic perspective, how these dynamics reflect and help illuminate the foundational level of temporal experience within which they are rooted, a path that Heidegger fails to explore.

For pragmatism, the beginning phase of scientific method, not as a formalized deductive model, not as a metaphysical enterprise illicitly reifying supposed ultimate truths, not as a causal analysis of humans and their environment,[3] but as lived experimental activity, exemplifies human creativity. Such scientific creativity arises out of the matrix of ordinary experience and in turn refers back to this everyday experience. The objects of systematic scientific creativity gain their fullness of meaning from the matrix of ordinary experience. Though the contents of an abstract scientific theory may be far removed from the qualitative aspects of everyday experience, such contents are not the found structures of some "ultimate reality of nature," but rather the possibility of their coming to be as objects of scientific knowledge requires and is founded upon the lived qualitative experience of the scientist. As Mead observes, "Controlled sensuous experience is the essential basis of all our science." Further, "the ultimate touchstone of reality is a piece of experience found in an unanalyzed world. . . . We can never retreat behind immediate experience to analyze elements that constitute the ultimate reality of all immediate experience, for whatever breath of reality these elements possess has been breathed into them by some unanalyzed experience."[4] In Dewey's terms, the refined products of scientific inquiry "inherit their full content of meaning within the context of actual experience."[5]

For Heidegger, the emergence of the scientific object requires a "working over" of the things in everyday experience, changing them to objects that are decontextualized from the involvements of the everyday world. The abstract objects of science that emerge from the operations of the scientist that work over the concrete things of everyday experience are not ontologically independent beings, but are dependent upon the everyday world of involvements from which they emerge through the creative activity of the scientist.[6]

Further, such creativity implies, for the pragmatist, a rejection of the "passive-spectator" view of knowledge and an introduction of the active, creative agent, who, through meanings, helps structure of objects of knowledge, and who cannot be separated from the world in which such objects emerge. As James well notes of scientific method, there is a big difference between verification, as the cause of the preservation of scientific conceptions, and creativity, as the cause of their production.[7] Dewey expresses this noetic creativity in science in his claim that "What is known is seen to be a product in which the act of observation plays a necessary role. Knowing is seen to be a participant in what is finally known." As he concludes, both perception and the meaningful backdrop within which it occurs are shot through with the interactional unity between knower and known.[8] Without such a unity there is no scientific world and there are no scientific objects.

As Heidegger observes, the opening of a scientific world, of a region, of a horizon within which a particular type of object can appear, is the fundamental event in research. Without the opening of a world of a particular type there are no scientific objects.[9] Thus, the projection of a scientific world determines the type of operations that can be used to yield objects within it. It determines the way in which the things of everyday experience are changed to yield the objects that conform to the possibilities allowed within the scientific world.[10] Projection provides the background from which specific types of possibilities emerge and within which particular types of objects can appear. This fundamental event in scientific research secures itself by proceeding rigorously according to its projected possibilities.[11]

For the pragmatist, such a creative noetic structuring of a world brings objects into an organizational focus from an indeterminate background and, as constitutive of meanings as dispositional modes of response, yields purposive, teleological or goal-oriented activity.[12] The system of meanings both sets the context for activity and rigorously limits the direction any activity takes, for such meaning structures are constituted by possibilities of acting toward a world.

This leads directly to Heidegger's understanding of methodology as the second essential characteristic of research.[13] Projection and the rigor it entails can develop as they do because of the characteristic of science that involves rule, law, explanation, and experiment, the various aspects of methodology. Rule and law bring objects into an organizational focus from a background of indeterminate change. "Only within the horizon of the incessant-otherness of change does the plenitude of particularity—of facts—show itself. But the facts must become objective. Hence procedure must represent the changeable in its changing, must bring it to a stand and let the motion be a motion nevertheless."[14] This is accomplished through rule and law. Rule allows for the "fixedness of facts and the constantness of their change," while "law" provides "the constancy of change in the necessity of its course."[15] A sphere of objects comes into representation through the establishing and verifying of rule and law. Experiment becomes possible with the laying down of rule and law as a basis, and explanation develops through experimentation. Explanation has a twofold function. "It accounts for an unknown by means of a known, and at the same time it verifies that known by means of that unknown."[16]

The latter aspect of the role of explanation leads directly to the final general characteristic of scientific method for pragmatism. The adequacy of meaning structures in grasping what is there, or in allowing what is there to reveal itself in a significant way, must be tested by consequences in experience. Only if the experiences anticipated by the possibilities of experience contained within the meaning structures are progressively fulfilled—though of course never completely and finally fulfilled—can truth be claimed for the assertions made. Initial feelings of assurance, initial insights, initial common assent, or any other origins of a theory do not determine its truth. Rather, to be counted as true, a claim must stand the test of the anticipated unfolding of experience. Such unfolding of experience in conformity with projected anticipations represents a self-corrective rather than a building-block model of knowledge. The meaning organizations or rules governing the organization of experiences are judged by their ability to turn a potentially indeterminate situation into a resolved or meaningfully experienced one. Thus Peirce stresses that scientific method is the only adequate method of fixing belief, for it is the only method by which beliefs must be tested and corrected by what experience presents.[17]

This self-corrective nature of the scientific enterprise leads to the final characterizations of the scientific endeavor as understood by Heidegger. Included in its ongoing activity is a drive to open up further possibilities for research.[18] Heidegger clearly distinguishes the depth of this self-corrective activity in the following claim:

> Ongoing activity becomes mere business whenever, in the pursuing of its methodology, it no longer keeps itself open on the basis of an ever-new accomplishing of its projection-plan, but only leaves that plan behind itself as a given; never again confirms and verifies its own self-accumulating results and the calculation of them, but simply chases after such results and calculations. Mere business must at all times be combated precisely because research is, in its essence, ongoing activity.[19]

Thus, the essential features of scientific method, as pure method, involve for both pragmatism and Heidegger the creative opening up in advance of a domain in which only things of a certain kind can show themselves. This world provides the context within which specific types of possibilities emerge as "there" for our consideration. In opening up a scientific domain, the scientist is casting a conceptual net by which to structure the emergence of facts within it. Only from this backdrop can the scientific endeavor determine which possibilities to utilize as it turns an indeterminate matrix into a realm of meaningful objects. Only through the opening of a world of a particular type can the purposive activity of the scientist proceed rigorously according to projected possibilities, testing explanations by their continuing ability to bring an indeterminate matrix into a network of meaningful objects through rule-guided activity. Finally, scientific method, as self-corrective ongoing activity, not only collects workable results within its present structure but also should reexamine the justification for the specific possibilities it has chosen for meaningful organization within the world of possibilities it has opened up through its initial projection. Neither Heidegger nor pragmatism characterize this latter aspect of ongoing activity as following clearly laid-down procedures, but rather tend to view it as involving a nonthematic, unarticulated grasp of the way to progress.

In light of this model of scientific method as pure method devoid of any particular type of content, the focus will turn to the context of the dynamics of everyday experience in pragmatism and in the philosophy of Heidegger to see what clarifying light it has to offer. This path is not followed by Heidegger, though a focus on his position from the direction of pragmatism can highlight its relevance for his understanding of the dynamics of concrete human existence.

The role of the function of scientific method in understanding everyday experience within pragmatic philosophy is evinced in several brief but telling remarks. As Dewey observes, awareness, even in its most primordial state, "represents a general trend of scientific inquiry." It means things entering, via directed activity, into a "condition of differential—or additive—change."[20] Or, as he summarizes, "There is no difference in kind between the methods of science and those of the plain man."[21] Peirce emphasizes the same point in his claim that the creative interpretations of scientific endeavor shade into everyday perceptual claims without any sharp line of demarcation between them.[22] Or, in Mead's terms, scientific method is embedded in the simplest process of perception of things in the world.[23] Again, Lewis attempts to clarify the noetic creativity ingredient in scientific objects by turning to the understanding of "thinghood" within common sense.[24]

Before turning to Heidegger's understanding of the world of lived experience that underlies science, it will be helpful to anticipate a few objections to the use within pragmatic philosophy of the above characterization of scientific method as the model for understanding the dynamics of all experience. One may object that, after all, scientific knowledge is theoretical knowledge and hence can surely not provide the model for understanding everyday experience. Is not the present analysis once again, though in a different manner, confusing the world of science with the world of lived experience? Has pragmatism, via a different route, again replaced the human mode of being with "technological being"? The answer here is a decisive "no." The use of the model is in no way an attempt to assert that perceptual experience is really a highly intellectual affair. Rather, the opposite is more the case. Scientific objects are highly sophisticated and intellectualized ways of dealing with experience at a second level, but they are not the product of any isolated intellect. Rather, the total concrete human way of being is involved in the very ordering of any level of awareness, and scientific knowledge partakes of the character of even the most rudimentary aspects by which a world of things emerges within experience.

Further, the scientific purpose of manipulation of the environment, and its use of scientific concepts as instruments of such manipulative control, are not technological maneuvers into which human activity is to be absorbed. Rather, again, the opposite is more the case. All activity constitutive of the human way of being, even at its most rudimentary level, is activity guided by direction in light of anticipatory rules that are meaningfully transformative of its environment. At more primordial levels of experience, such rules need not be, in fact are not, explicit or thematic, but rather are implicit in the anticipatory nature of the dynamic tendencies or potentialities and possibilities constitutive of the human way of being. As noted earlier, even at the level of scientific knowing there is an element of unthematic activity. Thus, the things of the everyday world, like the objects of science, are unified in terms of their function, not in terms of some underlying essence,[25] and the abstractly manipulative and instrumental purposes attributed to science have their roots at the foundation of the very possibility of human experience in general.

All experience incorporates the dynamics of scientific method, not in the sense that it is guided by sophisticated levels of thought, but in the sense that the very process that constitutes humans as dynamically developing, both in their way of knowing and their way of being, embodies the features revealed in the examination of scientific method. It is not that human experience, in any of its facets, is a lower or inferior form of scientific endeavor, but rather that scientific endeavor is a more explicit embodiment of the dynamics operative at all levels of experience, and hence the ingredients are more easy to distinguish. The pursuit of scientific knowledge is an endeavor throughout which are "writ large" the essential characters of any knowing, and it partakes of the character of even the most rudimentary ways in which human activity involves temporally extended anticipations of future experience to come.

The tendencies or potentialities constitutive of the human are "living meanings" that creatively structure the temporal present as it extends into the future, regulating the way in which things of various types can emerge within experience, setting the conditions of recognition for what will count as the coming to presence of instances of particular types of things. Such a processive structure of the human as a way of being and as a way of knowing are the conditions that make possible the emergence of the world of everyday active engagement as well as the scientific abstractions grounded in such an everyday world.[26] Just as the world of scientific objects is an abstraction from the rich matrix of everyday experience, so the world of the things of our everyday active engagement is there as the result of the context of meaningful projection through which our natural ontological embeddedness in a richer matrix enters into the contours of awareness.[27]

A proper understanding of the lessons of scientific method thus reveals that the nature into which the human is placed contains the qualitative fullness revealed in lived experience. In addition, the grasp of nature within the world is permeated with the meaning structures by which humans and their world are interactionally or intentionally bound, at the levels of both commonsense experience and scientific reflection. Pragmatism, in focusing on scientific method, is providing a phenomenologically or experientially based description of the lived-through activity of scientists that yields the emergence of their objects. In so doing, it is focusing on the explicit enlarged version of the conditions by which anything can emerge within experience, from the most rudimentary awareness of things within everyday experience to the most sophisticated objects of scientific knowledge. In providing a description of the lived experience within which the objects of science emerge, pragmatism is uncovering the essential aspects of the emergence of any contents of awareness. In brief, an examination of scientific method provides the features for understanding the very possibility of its existence as emerging from rudimentary experience.

In turning to Heidegger's philosophy it will be seen, in a similar way, that the difference between the levels of science and the everyday world of active engagement represents not a difference in method but a difference in type of content emerging from a common method. The lived world grounds the possibility of any scientific world. But the explicit, "enlarged version" of the dynamics of experience that emerges in understanding the method of science can in turn help clarify the method by which all contents of awareness emerge in their meaningfulness, though Heidegger does not pursue this path with its implications for temporality.

The previous analysis of Heidegger's understanding of scientific method showed that the abstract objects of a particular physical science are revealed in their technical physical structure to a person who is trained in that science; to a person who is operating according to that projection as the opening up of the possibilities of what kind of objects can appear in that physical world; to a person whose purposive activities proceed rigorously according to its laws and rules. For Heidegger, this projective activity of the scientist as scientist is similar to the projective activity of the ontological constitution of the human as human. What differs in science is the nature of the content that its projection allows to emerge, its level of abstractness, and the degree to which it is "freed" from the concreteness of everyday experience.

Although the projection that opens a world of modern physical science is mathematical, it is mathematical in this particular numerical way, according to Heidegger, only because it is mathematical in a broader sense. Mathematics, in its broader sense, means known in advance.[28] This knowing in advance is the result of a projection, regardless of how primordial and implicit such a projection may be. Further, though the rigor of mathematical physical science is exactitude, rigor is not identical with exactitude, for in the human sciences, rigor requires inexactitude.[29] Or, as he states, mathematics is not more rigorous than other disciplines, but more narrow.[30] Thus, its rigor can lie in exactitude. The rigorous adherence to the possibilities projected in original understanding manifests itself in the fact that anything there in our world emerges within an horizon of awareness prepared in advance.

For Heidegger, as discussed earlier, the human way of existing in a meaningful world is the context within which meaningful things can arise within experience. The horizon for a certain kind of experience must be projected in advance, must be opened up, before that kind of experience can take place. The objects of science do not disclose some underlying essence. Rather, they are "there" because of their function in a projected world, and this projected world of science can be there only because it and its objects are grounded in the everyday world. This everyday world is in turn grounded in the structural process of the human way of being as being-in-the-world, and within this everyday world the unities of things that emerge into presence lie analogously, for Heidegger, in their functionality, not in some underlying essence.

Any interpretive explanation is the development of the possibilities "there" because of an original projection. This original projection opens a realm within which possibilities emerge for consideration as possibilities to be seized upon. Even the most primordial grasp is not a presuppositionless apprehending of something that reveals itself, for the meaning of things is constituted, receives its significant structure from, our fore-having, fore-sight, and fore-conception.[31] Such interpretation may be implicit and concrete in our everyday concernful engagement with things in the world, or explicit and abstract in the scientific grasp of its world of objects. Each manner involves an interpretation of what it is to be, but with the level of science, "what it is to be" results from contents emerging through a narrowing, abstractive projection. There is, at all levels, an interplay of the constitution of meaning and the uncovering or unfolding or bringing forth of being. What emerges in world experience is an interactional unity of these two dimensions. Being is present within world experience, for world experience is the experience of being.[32]

Heidegger stresses that Roger Bacon cannot be the forerunner of modern science, for his method is that of observation rather than the opening up of a world; it represents the method of the spectator.[33] In the emergence of modern science, the modern era was oblivious to the non-spectator *method* that allowed its objects to emerge, and as a result reified its contents as absolute in spectator fashion. Only by understanding the method by which modern science achieved its contents can its foundations in, and continuity with, the dynamics of concrete experience be fully grasped and can it, in turn, serve to clarify such foundations.

The theoretical focus of science on its limited, narrow type of objects tends to conceal the very world of science that yields the possibility of these objects, as well as the projective emergence of this world from the world of everyday concrete existence. And, just as the object world of science tends to conceal the richer, less determinate matrix of the everyday world of active engagement from which it emerges, so the everyday world tends to conceal the richer, less determinate ontological grounding within which it emerges. The human engagement with things in the everyday world tends to conceal the world that gives these things its meaning, as well as the world-founding or world-projecting nature of the human mode of being that provides the regulated world structure within which a rich matrix of being can come into focus as an interrelated context of meaningful things. In pointing out the blindness of science in mistaking its content as the "absolutely there" for the human as spectator, in this way ignoring its world as the meaningful horizon emerging from its own world-founding activity, Heidegger in fact provides an "enlarged," explicit version of the way in which everyday being-in-the-world tends to be oblivious to world as the meaningful horizon and to world founding as the structure of the human way of being.

But, if Heidegger had actually utilized the model of scientific method in understanding concrete human existence, he might have noted another "covering over" regarding temporality in its cosmic implications. There was mention in the previous chapter that *Being and Time* still allows for the existence of substances other than the nonsubstantive human way of being, substances that ground the more abstract functional concepts.[34] In the later Heidegger the unity of things lies in their functionality.[35] But there is no recognition that functionality in experience relates to a broader context of temporality. This broader context, however, comes into view from the backdrop of the model of scientific method, for Heidegger holds that within science rule-guided behavior brings objects into an organizational focus from a background of indeterminate change. "Only within the horizon of the incessant otherness of change does the plenitude of particularity—of facts—show itself. But the facts must become objective. Hence procedure must represent the changeable in its changing, must bring it to a stand and let the motion be a motion nevertheless."[36]

Scientific method, as the model for understanding what is operative at the prescientific level, leads toward a metaphysics of process that is not that of scientific change within scientific time but a metaphysics of process that, as involving indeterminacy in its own way, allows us, as embedded in it, to "bring it to a stand" as the particularity of facts through rule-governed or anticipatory activity. If one utilizes the model of scientific method offered by Heidegger and pragmatism alike, there is no danger that the model of scientific method can lead to a "process metaphysics" of scientific change and scientific temporality, but rather it points the way toward a new understanding of cosmic temporality rooted in the dynamics of concrete human existence. If the significance of scientific method is not understood, however, such a potential danger can readily become a reality. Most appropriate here is Heidegger's own proclamation: "Where the danger is as the danger, there the saving power is already thriving also."[37] Pragmatic philosophy will be seen to be intent on definitively rescuing this "saving power" from the dangers in which it is embedded.

In retrospect, it should not be surprising that such diverse approaches as pragmatism and the philosophy of Heidegger tend toward a common vantage point. If, as phenomenology holds, there really is a pervasive structure to the human way of knowing, and any one instance can become the initial focal point for the interpretive descriptions that yield its pervasive features, then pragmatism can well begin with the dynamics of scientific knowing as an instance to be analyzed for its pervasive features and, with it as a model, proceed to understand the dynamics of concrete human existence. Conversely, if pragmatism is correct in its understanding of the method of science as yielding the pervasive features of all knowing, then such primordial knowing embedded in the dynamics of concrete human existence, as depicted by the descriptive interpretations of phenomenology, should include those features derived from the characteristics of scientific activity. And for both, these dynamics, which are at once descriptive of the human way of knowing and the human way of being, incorporate a creative, regulative structuring that renders an indeterminately rich processive matrix determinate for our anticipatory activity; renders an ongoing active, changing process "fixed" through the resultant emergence of objects as unities of function that we encounter, question, and explain in various ways and at various levels.

This regulative feature and the temporality in which it is embedded leads to the exploration in the next chapter of one further key point of contact between Heidegger and pragmatism rooted in their convergent understandings of time.

CHAPTER 5

Temporality and the A Priori: Heidegger and Pragmatism

It has been seen that Heidegger and pragmatism alike, with varying degrees of temporal implications, understand rule-guided behavior as the path through which an indeterminate nature enters our experience as the coherent, meaningful experience of things. This process in turn calls out for an a priori element within experience that regulates in advance the contours of this entry. Yet, with the exception of C. I. Lewis, the concept of an a priori is strangely lacking in pragmatic thought. Moreover, although Lewis refers to his understanding of the a priori as a pragmatic a priori, it not only by and large has not been utilized to throw light on the other pragmatists,[1] but is in fact one source of the general perception that Lewis does not quite fit in with these others. For his focus on the a priori is held to highlight a general concern with epistemic issues that distances him from the other pragmatists who are both less involved in his so-called abstract epistemic approach and display little interest in the a priori. For this very reason of a so-called abstract epistemic approach, a comparison of the nature and function of the a priori in the philosophies of Lewis and Heidegger may strike some as a strange endeavor, since Lewis's discussion arises within the context of his pervasive epistemic interests, while Heidegger's concern with the a priori, as with other issues, is deliberately ontological.

In rethinking Lewis's so-called abstract epistemic approach in relating it to Heidegger's position, the following examination will at the same time be expressing his position in a way that brings home its affinity with the other pragmatists. Thus, the present focus on Heidegger and Lewis can be understood more broadly as a focus on Heidegger and pragmatism in general, and the features of Lewis's philosophy that are here developed in some depth can be understood as features of the general position of classical American pragmatism as developed in the next part. Moreover, by bringing to light the shared novel characteristics of the a priori as it emerges within two diverse philosophical perspectives, each of which in its own way takes seriously the lived sense of temporal flow and its implications for issues of continuity and indeterminacy, this chapter is intended to highlight as well the significance of a novel "temporal a priori" for any view that proceeds along these same lines. With this broad backdrop in mind, the ensuing discussion will turn to the positions of Lewis and Heidegger.

The epistemology to which Heidegger and other phenomenologists rightly object is one to which Lewis objects as well, for they alike reject all remnants of Cartesian philosophy and the resultant epistemology that begins with a subject-object split, which illicitly detaches itself from the "external world" and then tries to prove its existence, attempting to put together again that which it never should have pulled apart.[2] Yet a rich and fundamental epistemological focus on existence can open meanings to the fullness and richness of their epistemic depth at the ground level of lived experience. At this level, the examination of the prereflective epistemic substrate for knowledge is inextricably intertwined with an explication of the structure and process of the being of the knower. In this way ontology and epistemology become two distinct perspectives from which reflection gets at the same foundational elements in existence. Thus Lewis, supposedly the most "epistemically" oriented of the pragmatic philosophers, understands knowing always within the context of being,[3] while Heidegger, in viewing understanding as an ontological structure of the human way of being, is concerned with focusing on understanding in a nonspectator fashion, as opposed to "the usual spectator sense of knowledge,"[4] thus casting his position in terms of traditionally epistemic alternatives.

Lewis's general pragmatic focus views meanings in the biological context of habits or attitudes of response.[5] This has unfortunately lead to far reaching misinterpretations of the position intended. As mentioned earlier, such a misinterpretation leads Heidegger to accuse pragmatism of being a scientific biologism.[6] However this pragmatic focus, in its richness, does not lie in opposition to a view of human awareness in terms of a field of meanings, but rather, when properly understood, reveals the purposive activity out of which awareness of meanings emerges. Human behavior is meaningful behavior, and it is in behavior that meaning is rooted. There is an inseparable relationship between the human biological organism bound to a natural environment and the perceiver who constitutes a world.[7] From the context of organic activity and behavioral environment there emerge irreducible meanings that allow a world of things to come to conscious awareness. Such meanings are irreducible to physical causal conditions or to psychological acts and processes, yet they emerge from the biological, when the biological is properly understood, for the content of human perception is inseparable from the structure of human behavior within its natural setting.

Purposive biological activity, as the foundation of meaning, then, cannot be understood in terms of scientific contents or scientific categories. Rather, it is the "lived through" biological activity of the human organism and, as such, is capable of phenomenological description. The phenomenological significance of habit is that such habits, dispositions, or tendencies are immediately experienced and pervade the very tone and structure of immediately grasped content.[8]

The purposive behavior that pervades and unifies perceptual awareness is precisely its binding intentional character as well. Hence Lewis can state that he is not advocating phenomenalism, but rather is presenting a "phenomenology of the perceptual."[9] Such an understanding of the rootedness of meanings in the context of a natural organism engrossed in a natural environment must be understood neither as biological nor psychological descriptions but as foundational, as an ontological description of the necessary conditions for the distinctively human experience of a perceived world.

Lewis holds that what humans encounter within their worldly experience is partially constituted by a system of meanings, rooted ultimately in behavior. Only through such meanings does that which is independent of humans reveal itself to them. He compares worldly facts to a landscape: "A landscape is a terrain, but a terrain as seeable by an eye. And a fact is a state of affairs, but a state of affairs as knowable by a mind."[10] In terms used in the previous chapters, that which is independent of human purposive activity enters into human purposive activity as the everyday worldly environment.

For Heidegger, as for Lewis, what is there independently of humans reveals itself as it becomes intraworldly, a point already discussed. That which is extant is not "other than" our everyday worldly environment, but is that which becomes our worldly environment as it enters the domain of the human. What is intraworldly reveals itself within the totality of our involvements with what is ready-to-hand, with what serves as "equipment" within the context of everyday purposive activity. This meaningful network of everyday involvements is the context of functionality that we call world and and that is preunderstood. This understanding is not an abstract knowing but is embedded in our purposive activity, our intentional orientation. Practical understanding is holistic. Being-in-the-world belongs to the being of the human as human, for we always intend some possibility of ourselves as a purpose. Human behavior is inherently purposive or intentional.

For Lewis and Heidegger alike, then, knowing emerges at a prereflective level of behavioral rapport with things in the world. For both there is a deepening and extending of intentionality, requiring a redefinition of understanding so that it now encompasses the whole intention and mode of existing of the human.[11] They both abandon theories according to which we experience brute sense data or qualities instead of things themselves that are encountered first through purposive activity. Because of this encounter through action Heidegger can claim that "the understanding of extantness is already implicit in the intentionality of perception, that is, in perceptual uncovering."[12] So also Lewis can assert, in analogous fashion, that independent reality "does not need to be assumed nor to be proved, but only to be acknowledged."[13] Moreover, as will be developed below, this view of purposive behavior or corporeal intentionality requires, for both, an understanding of time that rejects the traditional concept of the "now-point." Their shared understanding of what Heidegger calls "primordial time," the temporality that is a "reaching throughout" and that is more fundamental than the discrete moments or now-sequence of a derivative understanding of time, provides the basis for their shared understanding of the nature and function of the a priori.

Lewis uses the model of a mathematical rule generating a number series to illustrate the role of habit or purposive activity.[14] Two temporally rooted features emerge from this: the presence within experience of inexhaustible possibilities and the creatively regulative features of habit as anticipatory. Just as a mathematical rule may generate an unlimited series of numbers, so a disposition as a rule of organization contains within itself an unlimited number of possibilities for the generation of particular acts of response in relation to particular types of sensory content. Moreover, just as with the number series one may elicit any particular number, or any particular cut, but cannot exhaust via enumeration all possibilities, so one can elicit any particular set of sensory-action relations desired, but cannot exhaust via enumeration all possibilities. Furthermore, in none of these cases is the inability to exhaust via enumeration all possibilities a contingent fact, but rather is intrinsic to the nature of the generating rule. And as Lewis states, such an absence of boundedness gives rise to our "sense of the experientially possible but not experientially now actual."[15] Embodied in the actuality of our meaning structures as habits of response, then, is a sense of a reality that transcends actual occasions of experience.

A second and closely related feature of the model for understanding habit is that the rule for the generation of a series enters into the structure of that which is generated. The structure of that which is generated reflects the character of the rule that generates. The structure of human behavior as anticipatory both requires and makes possible the creatively regulative features of meaning as habit. The rulelike or regulative aspect of the a priori, as well as its emergence within the ongoing course of experience, is expressed in Dewey's claim that "A postulate as a rule of action is thus neither arbitrary nor externally a priori. It is not the former because it issues from the relation of means to the end to be reached. It is not the latter, because it is not imposed upon inquiry from without but is acknowledgment of that to which the undertaking of inquiry commits us."[16] Such a regulative rule is characterized by Dewey as "temporally a priori."[17]

The above features of habit are possible because, as Lewis so well expresses, "The absolutely given is a specious[18] present fading into the past and growing into the future with no genuine boundaries. The breaking of this up . . . marks already the activity of an interested mind."[19] The minimal experience always involves a durational flow, for it is filled with the rudimentary pulsations of the temporal structure of habit as anticipatory. The sense of the future involved in anticipatory activity is not an induction from past experience but is at the heart of experience in the durational present. Indeed, the experience of the "thick" durational present founds our very sense of the past. The sense of anticipatory activity thus incorporates the temporal directions of past, present, and future. Such a durational flow is essential for the pragmatic understanding of experience as experimental, for it involves an anticipation of a next experience to come, something for which we are waiting, an expectation set in motion by the temporal stretch of human activity. The temporally grounded structure of meaning as habit, then, is the source of the functionally organized concrete structural unity of perceived things as more than a collection of atomic appearances, and is the source of our sense of a reality whose possibilities of being experienced transcend all actual experiences.

For Heidegger, as for Lewis, to be an intentional agent such as we are is to be temporal and, as noted earlier, the unity of temporality comes into view as the structure of the human as such, with each dimension of temporality—past, present, and future, a "reaching throughout." The temporal dimensions of future, past, and present are represented in the temporal structure of purposeful action involving expecting, retaining, and making-present. 'Being' as such means making present. Our intentional comportments provide the prescriptive content that allows us access to the extant or independently real through worldly encounter in which things come to presence. The horizon for a certain kind of experience must be projected in advance, must be opened up, before that kind of experience can take place. Anticipation is ingredient in the fundamental nature of all understanding, and any interpretive explanation is the development of the possibilities "there" because of an original projection or meaning bestowing activity. This original projection opens a realm within which possibilities emerge for consideration as possibilities to be seized upon. Even the most primordial grasp is not a presuppositionless apprehending of something that reveals itself, for the meaning of things is constituted, receives its significant structure from, our fore-having, fore-sight, and fore-conception.[20]

For Lewis and Heidegger alike, then, the tendencies or potentialities constitutive of the human extend futurally forward and bring in toward themselves possibilities of experiencing consistent with these potentialities for structuring the temporal present as it extends into the future. Such potentialities, as creative and regulatory, as rules constitutive of the structural relationships within experience, are at once rules for the construction of the conditions within which things of various types can emerge within experience. Such conditions are not collections of actual or possible instances, but rather consist of the dynamically organized relational generalities that set the conditions of recognition for what will count as the emergence within experience of instances of particular types of things. Such orderings are resolutions to action embodied in the "living meaning" of the creative, regulative features of human behavior.

The temporal stretch of human experience as creative, regulative, and anticipatory reaches out to the future to let it emerge within the contours of the possibilities contained in the temporally rooted structure of meaning. Such a temporal structure of the human as a way of being and as a way of knowing is the condition that allows for the emergence of the world of everyday active engagement as well as the second-level reflections grounded in it, reflections that too often systematically deny the very temporal structure that founds their own possibility. This focus on temporality leads directly to Heidegger's understanding of schemata, which in turn forms an essential link with Lewis.[21]

According to Heidegger, schemata form the temporal horizons of encounter, and are products of the imagination rooted in primordial time. Primordial time is the basis of our finite knowledge and existence and is the source of the categories and the a priori as products of this temporal grounding in finitude.[22] The anticipatory structure of the schemata, grounded in primordial time, makes possible the objectivity of the object. This anticipatory structure is an orientation to receive, allowing for the meaningful encounter with anything given. This allows humans to reach out beyond the external aspects of the object to encounter it from within as an entity from the backdrop of a field. Schemata, in allowing for the objectivity of the object within the context of a field, thus provide the a priori conditions for knowledge.

The role of schemata in the nature of the a priori can best be further explored by turning to Heidegger's examination of Kant's examples of empirical and sensible concepts. In his creative appropriation of Kant, he focuses on the schema as a mode of sensibilization mediating the concept of understanding and image of sense. The schema is not an image, yet there is an image character to it, which Heidegger calls the "schema-image."[23] The schema is a representation of the rule that gives rise to it, and its possible schema-images contain a generality irreducible to any particular exemplification or series of exemplifications. For instance, the concept of 'dog' signifies a rule according to which my imagination can delineate the figure of a four-footed animal in a general manner, without limiting it to a determinate figure that any possible experience presents. The schema-image, then, as the image of the concept in its structural relation to the schema, mediates between the aspect of an object and a concept. The relation between schema-image and schema makes the schema-image a possible representation of the schematic rule of presentation. Hence, the concept is inseparable from and always refers directly to the schema, and beyond this regulative unity the concept is nothing.[24]

The image, then, derives its intuitive character from the "fact that the schema-image comes into being and from the way in which it comes into being from a possible representation which is represented in its regulative function, thus bringing the rule within the sphere of a possible intuition."[25] In this way, the rule is regulative for the possible intuition, and the image gains its intuitive character from the regulative rule and not from the brute content of sense. The schema-image as providing an aspect of an object, provides the possibility of an image. In Heidegger's words, "If, in general, a concept is that which serves as a rule, then the conceptual representation is the supplying, in advance, of the rule insofar as it provides an aspect corresponding to the specific way in which it regulates."[26] Or, he further specifies:

> The rule is represented in the *how* of its regulation, that is, according to the manner in which, in regulating the presentation, it inserts itself in, and imposes itself on, the aspect which presents the presentation. The act of representation of the *how* of the regulation is the free (i.e. not bound to a representation) "construction" (Bilden) of a sensibilization. The latter, in the sense just described, is the *source* of the image.[27]

From the backdrop of this analysis, the ensuing discussion can turn to Lewis's understanding of sense meaning as incorporating schematic imagery, and its role in understanding the a priori.

Lewis's focus on sense meaning has led to the popular interpretation of his position as putting forth a phenomenalistic reductionism that views meanings as reducible to the sense data out of which they are built.[28] But, though meaning is derivative from the sensuous, and though meanings themselves can be termed sensuous insofar as they refer to experience, meaning, even in its sensory aspect, cannot be reduced to the content of experience. The difference can perhaps best be clarified by stating that for Lewis the sensuous aspect of meaning provides, literally, the "sense" or principle or form by which humans interpret and organize the sensory aspect of experience. Meaning incorporates a response that, as an interpretive principle, enters into the very character of what is grasped, and as a result the meaning of any thing is irreducible to any data as existing apart from the character of the response. Only within the context of such an interpretive principle does the sensory come to awareness. Because of this, Lewis warns that there is a generally unnoticed complexity to sense meaning.[29]

Lewis introduces the schema in describing a sense meaning as a rule and imagined result.[30] An implicit sense meaning is a disposition or habit by which humans interact with the environment, while an explicit sense meaning is a precise "inspectable" schema or criterion in mind by which one grasps the presence of that to which a particular type of response is appropriate in order to gain a certain type of result.[31] As specifying types, the schema with its possible images or aspects is general as opposed to the particulars grasped by it. The schema image does not gain its intuitive character from the brute data of sense but rather it gains its intuitive structure "from above." Thus schemata, in providing for the possibility of the sensible grasp of the objectivity of the object, supply the a priori conditions for knowledge.

For neither Heidegger nor Lewis, then, can the schema-image be separated from the concept, but rather it is generated by, and gains its character from the regulative function of, the rule that generates it. Nor can the specific sensory content experienced be separated from the schema-image, for only within the possibilities presented by the schema-image can sensory content come to awareness. The regulative function of the rule is the source of the concrete unity of objectivity as more than a collection of appearances.

The positions of both Lewis and Heidegger can perhaps best be clarified by taking the term 'image' or 'schema-image' as 'aspect'. For example, one may say, quite correctly, that an ocean presents a turbulent image or aspect. And, while the specific empirical content of experience is best understood as one particular among many, the schema for the application of a living meaning or habit to experience is best understood as the one that determines the many. The importance of the content of the schematic structure lies in the way in which it comes into being. Such a structure represents an aspect of the dispositional structural order that regulates it, and that governs the possible transformations from one schematic aspect to another. Meaning as dispositional is a rule for the production of schematic aspects as the conditions for possible verifying instances. The living meaning virtually contains the conditions for its verification. Such conditions are not collections of actual or possible verifying instances, but rather consist of the relational generality of schematic aspects that set the conditions of recognition for what will count as verifying instances.

Further, for Lewis and Heidegger alike, a priori structures, as creative and regulatory of the way in which what exists independently of us can emerge within our experience, are rooted in intentional, temporally structured behavior and permeate perceptual awareness. Hence Heidegger asserts that categorial intuition[32] is just a concretion of the basic constitution of intentionality. Moreover, only on the basis of these can the a priori be understood.[33] This progression can be clarified by noting that "existential" structures or categories are found within a unity of reciprocity between perceiver and perceived made possible by the temporally founded structure of intentionality.[34] While the categories are not seen in perception as categories, these structures can be conceptually grasped as categories, and in this way a priori structures can be explored.

Thus Heidegger holds that the a priori is not subjective, but neither is it bound up with reality as extant, reality as it is before its entry into the horizons of our purposive activity. Rather the a priori structures the being of entities, the way beings *are* in our worldly encounter with them, the way the extant is brought to presence through purposeful activity.[35] He encapsulates this in his statement that "A priori means that which makes beings as beings possible in *what* and *how* they are."[36] In Lewis's words, the a priori within experience "exhibits the predetermined principles of interpretation, the criteria of our distinguishing and relating, of classification, and hence the criteria of reality of any sort."[37] For both Heidegger and Lewis, then, the a priori permeates perceptual awareness, manifesting itself in a unity of reciprocity between perceiver and perceived. Furthermore, it emerges through the temporally rooted anticipatory dimension of purposive or intentional activity that provides the possibilities that enable objects to come to awareness within the world or, in other terms, the possibilities that partially legislate the way objects can be within our world.

The a priori is prior, then, not because we recognize it earlier, or because it is formed prior to our contact with the sensible, but because possibilities in their possibility-making function have the character of the earlier in relation to that which they make possible in our worldly encounter.[38] This "earlier" is built into the temporal structure of our purposive or intentional behavior. In this way, the a priori has what Lewis calls a hypothetical necessity.[39] That is, if something is of a particular type, then certain features necessarily hold of it, for these features are constituted by the possibilities of experiencing contained in the purposive or intentional activity that allows it to emerge in experience as a particular type of thing. If the expectations are not verified, if the intention is not fulfilled, then it is not an object of that type. Alternative a priori structures will in turn give rise to alternative experiential possibilities. As Lewis stresses, the a priori has the seemingly paradoxical features of providing necessity yet having alternatives.[40] In a similar vein Peirce can claim that through a change of meanings, though not necessarily of words,[41] "what is inconceivable today may become conceivable tomorrow."[42]

Yet for Lewis, the a priori is analytic, while Heidegger speaks of an a priori synthesis. The answer here lies in further examining Lewis's intertwining of sense meaning and Kantian schemata and in distinguishing the temporal origin of the a priori from its structural nature.

Analytic truths, for Lewis, state relations between sense meanings and not merely between linguistic meanings. Lewis undercuts the conventionalist position that analytic truth expresses nothing beyond what is or can be determined by the language system that embodies it. Such a position makes the analyticity of a statement relative to the conventions of linguistic usage. To assert that the a priori is coextensive with the analytic, but dependent on linguistic conventions, fails, according to Lewis, to justify the epistemic function of the a priori. It explains the analytic by robbing it of its relevance to sense experience. Rather, the analyticity of the linguistic meaning is determined by the fixed intensional relationships of sense meanings.[43] To separate linguistic meaning from the sense meaning it conveys is to engage in a process of abstraction, for these two aspects are supplementary, not alternative, and "separable by abstraction rather than separated."[44] The abstraction of language from the sense meaning it expresses can serve a useful function for some purposes of analysis, for it is helpful to consider meaning in both its linguistic and in its sense aspects. However, Lewis considers it disastrous for philosophical theory when one makes the distinction absolute and posits linguistic meaning as the focal point for investigations. Such a procedure, by omitting the intensional relations among sense meanings, leads to the conventionalism that Lewis firmly rejects.

Heidegger stresses just this point in his objection to Mill's distinction between the analytic and the synthetic, the former being verbal. He argues that Mill's "allegedly verbal propositions cannot be completely severed from the beings they intend. . . . All significations, including those that are apparently mere verbal meanings, arise from reference to things. Every terminology presupposes some knowledge of things."[45] A priori structures are synthetic for Heidegger in that the necessary relationship is dependent upon the material meaning of the terms. But precisely this necessary reference to the sensory dimension, incorporated within the a priori in the form of schema and schema-image, provides the force of

Lewis's claim that the a priori is analytic, as will be seen below.

Lewis's theory of knowledge displays strong Kantian influences throughout. His position reveals both a sympathy toward yet sharp criticism of Kant's epistemology. This critical appreciation of the insights of Kant is nowhere more in evidence in Lewis's philosophy than in his rejection of Kant's synthetic a priori, a rejection based directly on Lewis's own use of the Kantian concept of schematism. As developed earlier, Lewis's theory of analyticity is rooted in the intensional relations of sense meaning, and these sense meanings are developed not as abstract images but as schemata in the Kantian sense. Using his concept of sense meaning as a schema for the application of a concept to experience, Lewis administers to Kant's theory of the synthetic a priori a neat, simple, yet forceful blow.

For Kant, the schematism imposes conditions not derivable from the conceptual definition, and hence a judgment relating the schematism to the concept is synthetic a priori. Lewis argues that this view of synthetic a priori judgments is plausible only if one overlooks an equivocation in Kant's argument, for the object referred to by the pure concept is a different object than the object to which the schematism is applicable. The pure concept, expressible by a definition and not limited to objects of possible experience, extends at least problematically to noumenal objects that need not be accessible to experience. The object, the essential criteria of which are given in the schematism, is a phenomenal object experienced in space and time. If we do not equivocate, but rather speak of the same phenomenal object in the subject and predicate, then the judgment is analytic. The concept of the phenomenal object entails all the spatial and temporal conditions essential to that object's being identifiable in experience. The schemata for the application of a concept to experience must be part of or contained within the generality of the meaning.[46]

Lewis offers a more brief, yet more general argument against the possibility of any synthetic a priori judgments. "Any character in the absence of which we should refuse to apply a term is of the essence. It is included in the signification of the term; and, any definition which does not entail such an essential character represents a faulty analysis of the meaning in question."[47] Such faulty analyses are readily understandable, for the intrinsic intensional connections among terms are rooted in sense meanings, and our verbal expression of these meanings may well omit too much of the concrete ground of the connection. Through this failure of analysis, the appearance of synthetic a priori judgments can arise. But, if the analysis of a meaning is adequate, the schema is included in the statement of the real definition. And that which is part of the definition of a term is analytically related to it.

Both Lewis's theory of meaning and his notion of the test schema reveal that his theory of analyticity is basically one of inclusion or containment, not synonymy. Through the logic of the functioning of sense meaning as dispositional Lewis can provide a solution to the problem of containment, for this functioning explains the sense in which one can discover that some quality or character not explicit in a definition is nevertheless essential to the meaning in question. Similarly, this functioning of sense meaning makes clear the way in which a meaning of which one is not conscious may be implicit in that meaning of which one is conscious. The answer lies in the distinction between the concrete disposition or habit as the rule of organization and the awareness of the schematic aspects organized by the rule. Peirce likewise finds the answer in this distinction,[48] and as a result can state that the living meaning "virtually contains" these aspects.[49] In brief, a habit or disposition is a "living meaning" that generates acts of response in relation to criteria for grasping the situations in terms of which such activity is appropriate. Such a "living meaning" is inspectable in any aspect through its possible particular applications, but can never be reduced to any series of such applications. Turning again to the analogy of a mathematical rule generating a number series, it can be said that in understanding the relational structure of that which is generated, we understand to that extent the rule that generates it.

A priori rules, rooted in dispositional tendencies, then, legislate ways in which experience can be interpreted, and a priori truths explicate the implicational relations contained within and among them. Genetically, these rules arise though the cumulative effect of past experience and the creative synthesis or fixation, within the ongoing course of experience, of dispositionally organized relationships among possible experiences. The fixation intended corresponds most closely to the creative process that Peirce calls abduction, though what are here fixed by such creative activity or abductive processes are not empirical hypotheses but the meaningful structures within which we can begin to form empirical hypotheses. And habit, as creatively structuring, always brings a "more than" to the organization of past experience. It is at this point that the creative synthesis functions in relation to "material meanings." But, at any point in the experiential process, the dispositional meaning logically contains all that it has creatively synthesized or "fixated," or conversely, all that it now has the power or potential to generate.

While the genesis of an a priori structure involves a creative synthesis, then, its structural nature, which is rooted in the absoluteness of relationships grounded in sense meanings, is analytic. Dewey indicates the key role of an a priori regulative feature that is distinctively analytic in his assertion that "All prepositions of existential import involve delimiting analytic operations of observation. . . . The operations of observation executed are controlled by conceptions which define the conditions to be satisfied . . . in descriptive determination of kinds."[50] Such intensionally grounded analytic relationships emerging from purposive activity pervade all levels of worldly experience, from the most rudimentary expectations of prereflective experience to sophisticated scientific knowledge and the development of abstract formal systems.[51] Important meaning structures may be difficult to capture by tracing which concepts are included in which in an explicit articulation of a relationship. But explicit articulation of the analysis of a concept is ultimately an attempt to capture what has been implicitly operative in the structure of our purposive activity, a structure that contains the schematic forms of its applicability.

James's refusal to relate issues of the a priori to the analytic is instructive of his understanding of the a priori. He holds that the analytic-synthetic distinction is an unhappy legacy from Kant that loses all philosophical interest the moment one ceases to ascribe to any a priori truths that "legislative character for all possible experience" that Kant believed in.[52] "We ourselves have denied such legislative character, and contended that it was for experience itself to prove whether its data can or cannot be assimilated to those ideal terms between which a priori relations obtain. The analytic-synthetic debate is thus for us devoid of all significance."[53] James thinks that the experimental context of a priori truths—the temporal origin of a priori relationships and the testing of their usefulness in terms of their workability in experience—makes the issue of the analytic-synthetic distinction useless, since there is no claim for any a priori truth legislative of all experience in Kant's sense.[54] James seems in fact to be groping toward a position that implies not a rejection of the entire analytic-synthetic issue but a reconstruction of the entire nature of the issue.

Within such a reconstructed context, the claims implicit in conceptual interrelationships cannot be reduced to the contents of experience or to empirical generalizations about relationships within experience, for they are creatively synthesized legislative rules for what types of objects or facts can conceivably emerge within experience to provide the basis for empirical generalizations about facts and objects. Further, they are legislative for the very possibility of the emergence of anything within experience. And, though a priori legislative for future experience and containing analytic, intensionally grounded relationships, yet their justification is ultimately pragmatic or functional: if the realities of various types that they delimit do not adequately work in capturing the indeterminate richness of experience, then they will be replaced by logically distinct and new meanings, new intentions.

This understanding of the a priori is inextricably woven into Lewis's subtly complex understanding of a temporally founded noetic creativity that is rooted in purposive biological activity and constitutive of the nature of experience as experimental. A similar understanding of the a priori is also inextricably woven into Heidegger's phenomenological grounding of the a priori in temporally rooted finitude. The meaning of any intending, the character of its possibilities of fulfillment in experience, and the character of the horizon in which it functions are all mutually interdependent, and all are embodied in concrete purposive activity founded in the finitude of temporal existence and alterable within its flow.

For Heidegger and Lewis alike, then, the a priori element in experience, which regulates in advance the possibility of the emergence within experience of facts and objects, is rooted in the concreteness of rudimentary purposive or intentional human behavior, and it arises from, is made possible by, and is replaceable or alterable within, the finite temporal structure of such behavior. Further, while it emerges through a behaviorally operative creative synthesis, it then contains within its structure the conditions for its applicability in the ongoing course of experience. This containment houses within its very structure a reference to the sensible, and hence the features of its relational structure can be apprehended within sensible experience. This a priori element operative in human behavior and incorporated in the very shaping of a world makes possible a perspectival grasp of an independent processive universe. In so doing, it provides the vehicle by which we render determinate within our world the inherent indeterminacy of an unfolding temporal universe in which its indefinite richness of possibilities and potentialities reach throughout past, present, and future.

These various characteristics of the a priori operative in the philosophies of Heidegger and Lewis, which emerge from a shared understanding of a foundational behavioral level rooted ultimately in their shared understanding of temporality and its significance, in an important sense undercut some of the contemporary debates about the nature of analytic truths as articulated in explicit claims. More significantly, this understanding of the a priori changes as well the very nature of the analytic-synthetic debate, casting each of the "opposing alternatives" in a new, fruitful, and complementary light. It carries the analytic-synthetic distinction beyond the conventionalism of language, beyond the functionalism of decisions of usage, and beyond the distinction between logic and knowledge about the world. It founds the absoluteness of the analyticity of an a priori that coerces the mind within any system but that nonetheless can be exchanged for another system, and that is linked to experience by the method of experimentation; for it founds an a priori analyticity that both arises from the concretely rich matrix of experience and is justified by the intelligibility it introduces into experience.

Had Heidegger been attuned to the way in which scientific method not only arises from primordial experience but also provides the model for understanding the dynamics of such experience, he may have more openly focused on the contextual, experimental, and pragmatically justifiable dimensions of the a priori that are inherently embedded in its operation through its radically temporal nature. It is this broadly based reformulation of the a priori by Heidegger and Lewis alike that the present chapter has endeavored to abstract out from its embeddedness in diverse historical contexts, with the hope both that their respective expressions of this reformulation can each help illuminate the other and that this mutual illumination can in turn help highlight the far-ranging significance of a shared understanding of time. While those who hold to the key role of interpretation in an ontologically dense, vague universe consisting of discrete units could find utilization for such a temporal a priori as elicited here, yet the development of this notion of the a priori is dependent upon the functioning of habits as concrete dispositional tendencies and, as argued in part I, the problem of the ultimacy of discrete units of time renders problematic the possible functioning of such continuities.

The next part, in focusing more directly on the development of a pragmatic framework for temporality and related issues, will first turn to a reconstruction of Mead's understanding of the self in order to rid his position of implicit remnants of temporal discreteness and the resulting worrisome implications to which this gives rise, thus paving the way for a deepened understanding of the role of the self in the temporal reintegration of humans and nature.

PART III

The Pragmatic Reconstruction of Time

CHAPTER 6

The Temporal Self: A Restructuring of Mead and a Lesson in the Need for Vigilance

Any understanding of the self and the way its distinctiveness from all other objects is marked out is inseparably interrelated with the way its temporal experience is characterized. Whithehead's understanding of actual occasions and prehensions is found inadequte by Neville precisely in the area of personal identity through time, for "a personal agent as a complex society of actual occasions each prehending and hence including its predecessors, at least with respect to important reiterations of pattern, does not adequately mark off the distinction between the aggregative quality of continuous human identity and the aggregative quality of the environment in which the person is embedded."[1] Moreover, as discussed in chapter 2, Derrida's dismantling of the self shares the consequences of his fragmented view of time.

Later chapters will show that Mead accepts the pragmatic position that lived temporality is partially constituted by and constitutive of human praxis, incorporates a temporally extended present within which experience opens onto past and future and to which past and future adjust, and involves the direct sense of change and temporal direction. Yet this very sense of the temporal, corporeal existence of myself seems in various ways oddly ignored or oddly distorted in his understanding of the I-me cognitive dynamics constitutive of selfhood, giving rise in Mead's discussion of the self to a temporal tension similar to that operative in Husserl's position on time and exploited in Derrida's deconstruction of presence and self. Indeed, the temporal basis for Derrida's loss of the self is implicitly operative as one pole of a tension in Mead's understanding of the self. This retrogression, which implicitly intrudes into Mead's understanding of the self, has itself gone strangely unnoticed by his critics. This is perhaps explained by the focus Mead scholars tend to bring to his philosophy.

G. H. Mead has possibly the most developed view of both the self and of time in American philosophy. Yet, while his view of the self is widely discussed by philosophers, psychologists, and sociologists, among others, his view of time is relegated usually to the background, often to oblivion. But it is in fact the unrecognized source of their numerous and usually frustrating searches for the "I" as active agent, searches that turn up a "me" quite readily, but leave the "I" shrouded in varying degrees of mystery, at times leading these critics to the conclusion that the self is identical with the "me," that the "I" is a fiction.

This chapter will explore Mead's understanding of the self as cognitive to show the temporal tensions that are operative within it and illustrate the problems to which this leads. It will also utilize certain incipient tendencies within his position to reconstruct it in a way that offers a deepened and expanded understanding of the self in terms of its temporal dimensions and its embeddedness in a temporal universe.

For Mead, the "I" and the "me" aspects of the self are neither metaphysically nor numerically distinct but rather are functional distinctions that, in their inseparable interrelation, constitute the self. The "I" is the functional pole or functional dimension of the self as the immediacy of the present, spontaneity, creativity, the individual perspective, the subject pole. The "me" is the functional dimension of the self that represents the constraints of the past, of tradition, of culture and institutionalized practices, the shaping of the self by the community perspective or what Mead calls the generalized other. It also represents the object pole. Mead's stress on the "me" and on the elusiveness of the "I" throughout his writings is intended to avoid the psychological and philosophical fallacies of introducing a supposed experience of sensations representing an objective reality as well as a supposed reflective or cognitive or intuitive grasp of the immediacy of an "I." The "I" as subject can never exist as an object for consciousness. The "I" by its very *function* can never be an object of reflection, because in being reflected upon, it becomes a "me."

Mead admits that there is, as a "more or less constant feature of our consciousness," a "running current of awareness of what we do," which has often led to the assumption of a self that is conscious of both subject and object, a subject acting toward an object world and at the same time directly conscious of this subject as subject. His response here is that the observer of our self-conscious conduct is not the "I" but the rapidly changing series of "me's" that occur in experience.[2] In a related but slightly different vein, he recognizes that there is a way of observing oneself in which both the observer and the observed do appear. For example, a person remembers asking himself about something, scolding himself for his actions, congratulating himself for his achievements, and so on. Though one finds both a subject and an object, however, "it is a subject that is now an object of observation and has the same nature as the object self."[3] In brief, it is not a true "I," but a "me." No reflection can ever yield a true "I," for reflection can only yield an *object of* reflection.[4]

Only in memory is the "I" constantly present in experience according to Mead. As he stresses, "the 'I' of this moment is present in the 'me' of the next moment. There again I cannot turn around quick enough to catch myself. I become a 'me,' in so far as I remember what I said. . . . The 'I' in memory is there as the spokesman of the self of the second, or minute or day ago. As given, it is a 'me,' but it is a 'me' which was the 'I' at the earlier time."[5] As Mead stresses, "When the self becomes an object, it appears in memory."[6] Thus, he holds that the "I" enters one's experience as an historical figure, and while a "me" is inconceivable without an "I," such an "I" must be a presupposition, for it is never "there" as a presentation of conscious experience[7] (p. 142).

Mead concludes that "the very conversational character of our inner experience, the very process of replying to one's own talk, implies an 'I' behind the scenes who answers to the gestures, the symbols that arise in consciousness."[8] As behind the scenes, this "I" is, in Mead's words, like "the transcendental self of Kant, the soul that James conceived behind the scene holding on to the skirts of an idea to give it an added increment of emphasis."[9] Thus Mead claims, in extreme form, that the "I" is "fictitious" in that it is always out of sight of itself.[10]

Yet the creative activity of the "I" pole is not really "behind the scene" but "within the scene." Mead points toward this in various ways but backs off, seemingly always afraid of the reappearance of a running current of psychical awareness paralleling activity—either as an intuited "I" or as representative contents, to which he so strongly objects. Perhaps this concern explains to some extent his failure to adequately explore the implications of his own analyses of concrete, corporeal temporal existence for the nature of self-awareness, for such an exploration will be shown to in fact place the "I" in a certain sense within the scene of the present. To further develop this point, the ensuing discussion will turn to a feature of the self that is demanded by Mead's focus on human action but that is so often and so oddly overlooked by both himself and the many commentators on his position: that is, the functioning of the corporeal or "existential" subject.

This overlooked ontological grounding of the cognitive or dialogical nature of the self in the lived body rooted in time, comes into focus if one takes seriously Mead's claim that the distinction between the "I" and the "me" is functional in that it is made "from the point of view of conduct itself."[11] The "I" and the "me" represent "two distinguishable phases"[12] of the ongoing conduct of a concrete organism. The "I" and the "me" emerge as two dimensions of the concrete subject that, in its vital functioning in the temporal flow, provides the ontological grounding for the sense of the cognitive or dialogical stretch. Here it is important to distinguish between ambiguous uses in Mead's writings of "the body." Mead holds that "we can distinguish very definitely between the self and the body. . . . It is perfectly true that the eye can see the foot, but it does not see the body as a whole. . . . The body does not experience itself as a whole, in the sense in which the self in some way enters into the experience of the self."[13]

The existential sense of the lived body, however, is not something partially grasped by sight, the body as something that I have. Rather it is something that is a lived unity grasped as a whole in experience, the body that I am, the body that is an expression of selfhood. Mead implicitly points toward this when he stressses that the self "must be related to the entire body."[14] His ambiguous focus on the body in these two different ways leads to his diverse treatments of the body in relation to the self. And it is the latter brief, scattered, and rare treatment of the body that serves as a fruitful point of departure for understanding the role of the lived body, the body that I am, in revealing the inextricable intertwining of the experiential sense of time with the cosmic time of the universe in which we are embedded.

Mead holds that the social aspect of the self, "the me," the incorporation of the generalized other, arises before the "I" aspect can emerge. One has an "I" only over against or in relation to the "me." The "I" pole emerges subsequent to the "me" pole as its correlate, and the internal conversation between the "I," and the "me" as the voice of the other, becomes possible. As Mead carefully points out, however, though the voice is the voice of another, the source of it all is one's self—the organized group of impulses that he calls the biologic individual or biologic self.[15] The term 'biologic' is used by him to emphasize the living reality as distinguished from reflection.[16] Moreover, this biologic individual or self adjusted to its social environment, and through this to the world at large, is what is grasped as an object in self-reflection.[17]

The term 'biologic individual' or 'biological self' can be misleading in bringing to mind shades of reductionism. The term 'lived body' or 'body self' will thus be used instead in developing the implications of this dimension. This is the self in its primordial biological dimension, the concrete, decentered subject in its living reality as the source of reflective awareness and hence of the I-me poles of internal dialogue. It is prior to and the source of the cognitive I-me distinction, prior to its appropriation of itself as an object of reflection, as both knower and known. It is the lived body as purposive and social, the concrete decentered subject that is prior to reflection but that, once reflection has arisen, is appropriated as a "me" by an "I."

To view the *source of* the cognitive relation as in any sense a self, even a body-self, may appear strange for a position using Mead as a point of departure, for the self, for Mead, is essentially cognitive, consisting of the nature of a dialogue between the "I" and the "me." Yet in a way this is most appropriate, for the functioning of the concrete subject that gives rise to the I-me cognitive dynamics of the self is not discarded when the reflective self arises in experience. It is never discarded but subtends such dynamics; it is not just genetically prior to the cognitive self but logically prior in its ongoing dynamics. The concrete subject, as that which underlies the I-me distinction, is a dimension of the self, the "living reality" that pervades reflection, and it incorporates the features of present creativity passing into a more settled past. This biological dimension is a lived unity grasped as a whole in experience, the body that I am, the body that is an expression of selfhood. While Mead clearly recognizes that "The self is not so much a substance as a process in which the conversation of gestures has been internalized within an organic form,"[18] yet the discussion of the self becomes detatched from the temporal significance of this organic form and its incorporation into the very nature of selfhood.

The importance of this dimension can be gleaned from Mead's claims concerning what he considers "one of the problems of most of our conscious experience," that is, the getting into experience of the "I" with which we do, in a certain sense, identify ourselves.[19] He holds that one cannot get the immediate response of the "I" in this ongoing process because, though the organism brings parts of itself into its environment, it does not "bring the life-process itself into the environment."[20] To bring the "I" into experience, then, is to get hold of the life process, the living reality of the concrete subject, in some way. But ultimately his misappropriation of his own understanding of time rules out this possibility.

An example that Mead gives as an analogy for self-awareness is most revealing of his hesitancy to follow through for his analysis of the self the implications of his own analysis of temporality, for in giving an example that seems made to order to highlight the direct sense of change, the direct sense of the process of the passage of events, he ignores this aspect. In the example he points out that "We sense a ball falling as it passes, and as it does pass part of the ball is covered and part is being uncovered. We remember where the ball was a moment ago and we anticipate where it will be beyond what is given in our experience."[21] Mead concludes that the same holds of ourselves, for to look back and see what we are doing involves getting a memory.[22] However, what the falling ball analogy clearly points out is that in addition to remembering where the ball was a moment ago, there is a direct sense of the falling ball as it passes, a direct sense of part of the ball in the process of being covered and part in the process of being uncovered, a direct sense of change that is incorporated into the sense of the passing present. Mead exemplifies this elsewhere, in a discussion not focused on selfhood, through his analysis of the flight of a bird: "There is not only the flight of a bird before the individual but also the marking to the individual of the separate positions of the bird within his so-called apperceptive grasp. Such a set of indications to himself . . . marks the limits of the immediate change in his experience."[23]

This direct sense of change in the passing present is "there," and in relation to the self would appear to be a sense of the passing from the "I" to the "me." This direct sense of change ultimately illustrated by both analogies, rather than the memory upon which he focuses, best explains why Mead can hold that the falling ball analogy shows how, on the basis of "experience we distinguish that individual who is doing something from the 'me' who puts the problem up to him."[24] That sense of movement is "there" in the present, an I-me dynamic in an ongoing, continuous development, an I-me dialogue that is rooted in, and evinces the temporal vitality of, the concrete subject, and that is a continual uncovering and covering of its ongoing development. Thus, it is true that the "'I' does not appear *in the same sense* in experience as does the "me."[25] The "I," as Mead states, "does not get into the limelight,"[26] but neither is it hiding behind stage. To look back and see what we are doing does, as Mead states, involve memory, yet the sense of the ongoing I-me dynamics, though not seen, though not an object of knowledge, is yet precisely that, sensed.

To clarify this point further, the following development will leave Mead's falling ball analogy and turn to his general concept of sociality[27] as the stage of transition or phase of adjustment between the arising of the novel within the old order and the reoganization that gives rise to the new. The term 'sociality' refers not to the new order or to the old order, but to the process of adjustment in which the novel occurrence belongs both to the old order that provides the conditions for its emergence and the new order that accommodates it and renders it continuous with what came before. Sociality is the very character of the present, and its temporal character is essential to the nature of the self.

Applying this concept of sociality as constitutive of the present to the issue at hand, the "I," as representing the present, represents as well the emerging novelty of the present. The novel response of the "I" is the emergent event, the conditions for which are provided by the past "me" system, and whose coming to be requires a new "me" system that comes to terms with the emergent event by incorporating it. In this period of adjustment, the novel, which emerges as novel out of the old system, is accommodated by the new system that, in its adjustment, reinstates it as continuous with what came before. The novel response of the "I" in the present emerges from the old me, and as an emergent could not be predicted from the old me. To this emerging novelty a new me must adjust. The change from the old me system to the new me system is brought about by a process of adjustment in the present, and the transition from the old "me" to the new "me" via the emergence of the novelty of the "I," a process of adjustment involving the ontologically grounded temporal vitality of the concrete subject, is immediately experienced, in its covering and uncovering, through the experience of change in the passing present. Reflection is rooted in the ontologically grounded temporality of the lived body. An adequate account of the reflection constitutive of selfhoof therefore must incorporate the features of temporality.

As already mentioned, any "me" contains attitudes partially shaped by the past creativity of an "I," for the previous attitudes or actions of the "I" have changed the generalized other even as the generalized other has influenced the actions or attitudes of the "I." The "me" is an inseparable fusion and funding of the historically changing, developing circumstances or social contexts in which one finds oneself, and one's own ongoing unique responses that both alter the social dimension and make its incorporation into the "me" uniquely mine. The sense of the object self as a whole represents not just conformity or stability but past creativity. But the sense of creativity that once was is not possible without the sense of creativity, and creativity is located in a passing present. H. S. Thayer points out that for Mead the "I" as the creative pole in internal dialogue has its very genesis in memory,[28] but, again, there is perhaps some difficulty in remembering what one has never experienced. In the sense of the emerging novelty of the "I" and the accompanying adjustment of the "me," the sense of one's creativity passes into the sense of having once been.

What the above indicates is that not only is the "I" not hiding behind the stage of the present, but neither can the "me" be pushed from it. Mead illicitly conflates the temporal dimensions of the "me" by confounding the "me" as a pole in the dialogical stretch of the temporal present with the "me" as an earlier self thematized in memory through a "second-level" self-reflection. This happens, at least in part, because he illicitly separates or "compartmentalizes" his insights concerning the dialogical nature of the self and his perceptive analyses of temporality. Such a separation cannot be made, for the dialogical nature of the self is ontologically grounded, via the lived body, in the dimensions of sociality that constitute the nature of time.[29]

The fundamental concern of reflection is not to catch the "I" as an object of thematized reflective awareness—which cannot be done, not to engage in retrospection, remembering what one said or did or thought, but to carry on an inner dialogue and to anticipate and critically evaluate its intended probable results. The inner "I-me" dialogue as it operates in a passing present funded with possibilities emerging from a past and oriented toward a future is nonthematic in that the reflexivity is not thematically aware of itself. The "I" addresses and responds to the "me" as object, but without being thematically aware of its own reflexivity, though it can have a *sense* of reflexivity *in the process of* bending back upon itself. In the active engagement in internal dialogue taking place in the passing present, there is a sense of the covering and uncovering, a sense of the "dialogical stretch" of the temporal present. This sense of the cognitive dialogical stretch is ontologically grounded in the temporal dynamics of the concrete, decentered subject and subtends any cognitive or dialogical "I-me" relation.

Mead's analysis of time, when related to the self, points to the need for a focus on the temporal import of the distinction between retention and memory in a way similar to that found in Husserl and attacked by Derrida. Mead stresses that the self, as an object found in memory, can be an object found through "immediate memory." Immediate memory, however, is very different from the direct grasp of the immediate past as it is *present within* the passing present. Retention in a passing present is not memory; not even immediate memory. The dialogical stretch of the self in the passing present involves retention. In this passing present there is at once a direct sense of the past and the future and of the move from one toward the other that underlies both memory and expectation. As will be developed further in later chapaters, time is not a "thin" linear direction, but rather the temporal layers of the past are ingredient in the funded present, thus entering into the retentions of the present involved in the dialogical stretch, a stretch ontologically grounded in the ongoing temporal dynamics of the lived body with its dimensions of sociality.

The nonthematic sense of self-functioning is not caught in the problematics of psychical awareness. The understanding of the sense of the cognitive self as a sense ultimately rooted in the ongoing temporal dynamics of the concrete or decentered subject renders such a problem spurious. The ongoing nonthematic sense of self does not parallel experience in any way but infuses experience with the living dynamics through which experience is "my experience." What is sensed in experience is thus not an awareness running parallel to ongoing activity, but the life process of the ongoing function of the creativity of the present passing into the more settled character of the past, and the temporal vitality of this process permeates the "I"-me" dialogue of cognitive activity, of corporeal intentionality in the process of bending back upon itself, providing the running current of awareness that Mead attempts to explain in terms of a series of rapidly changing "me's" of moments immediately past.

Though Mead recognizes the self as inherently temporal, as incorporating a past oriented toward a future, yet his discussion of the "I-me" distinction is not infrequently couched in peculiarly atomistic terms, a tendency implicit in his discussion of the rapidly changing series of "me's" in terms of which he attempts to exhaust the running current of awareness, to explain the awareness of self as always an awareness of "the moment passed,"[30] but never of the passing. This problem is highlighted in Mead's discussion of the child's assumption of diverse roles. Though Mead stresses that even "The simplest form of being another to one's self . . . involves a temporal situation,"[31] this temporal situation takes a strange twist, for he claims that "It is in this field that the continuous flow breaks up in ordered series. . . . Time with its distinguishable moments enters, so to speak, with the intervals necessary to shift the scene and change the costumes. *One cannot be another and yet himself except from the standpoint of a time which is composed of entirely independent elements.*"[32]

What this statement appears to indicate, taken in conjunction with Mead's perceptive analyses of temporality, is that the very constitution of the self is distortive of the true nature of temporality. Mead, in his discussion of the self in terms of the "I" and the "me," seems inadvertently to have shifted the scene and changed the costumes of time, as a result dealing with the "I" as the locus of a momentary present detached from the direct sense of change and from the dynamics of sociality that constitute all temporal presents. This relapse allows for an invisible "I" that catches a past "me" in reflective activity characterized by discrete moments, but does not allow for an "I" that, as representing the emerging creativity of the concrete subject in a temporally extended present characterized by the features of sociality, is experienced in its "covering and uncovering" in the span of adjustment in which it is in the process of passing from a "me" that provided the conditions for its novel activity to a "me" that comes to terms with the novelty and renders it continuous with what came before.

Mead's tendency to ignore the features of temporality that his own analysis has revealed is perhaps motivated by his concern to avoid all remnants of psychical contents or of an intuited "I." Yet, as previously stressed, his understanding of temporality undercuts the very possibility of the types of positions to which he objects. The nonthematic awareness of the functioning constitutive of selfhood does not deny the outward directedness of direct awareness, does not allow for an intuited "I" or for psychical contents over against an objectively perceived world or self, but for a sense of the decentered corporeal subject and the cognitive dynamics it grounds.

Mead's ambivalence concerning the relation of self and temporalty can be found in the following claims, which at first glance are similar but which have radically different implications. He holds that "We are aware of ourselves, and of what the situation is, but exactly how we will act never gets into experience until *after* the action takes place.[33] Yet, in dealing with the self as a structure in terms of process, he also states that "We are finding out what we are going to say, what we are going to do, by saying and doing, and *in the process* we are continually controlling the process itself.[34] In contrast to the former claim, this latter statement, as well as his entire discussion of temporality, seemingly indicates that the action gets into our experience not *after*, but *as*, it is taking place. This view allows Mead to say that as the response taking place appears in experience it is *not* totally as memory experience.[35] Thinking itself lies "betwixt and between the old system and the new," and in the "betwixt and between," which constitutes the present, one does not experience just a series of remembered "me's," but the passage from an old "me" to a new "me" via the novel activity of the "I."

The roots of this ambivalence in Mead's thinking concerning the nature of the self in relation to temporality can perhaps be traced to his dual uses of sociality and his characterization of each in terms of temporality.[36] Mead uses social or sociality in the intrasystematic sense of being with other objects or organisms in the same system and in the intersystematic sense of being in two systems at once. The self must of course be understood as social in both senses. As Mead characterizes the distinction in terms of temporality, however, the second sense of sociality involves passage, while the first represents a system as it can conceivably be taken at an instant: As he states:

> The difference between these two dimensions of sociality is temporal. A system can conceivably be taken at an instant, and the social character of the individual member would in that instant be what it is because of the mutual relationships of all members. On the other hand, an object can be a member of two divergent systems only in passage, in which its nature in one system leads to the transformation which its passing into another system carries with it. In the passage itself it can be in both.[37]

Mead, in his discussion of the self, more than not deals with the self in terms of the abstraction of instants, forgetting that any discussion of the self that draws from this view of time is ultimately an abstraction from the sense of lived temporal passage.

The living dynamics of the decentered subject, the biological or existential dimension of the self that can function both as an "I" and a "me," is at once both and neither. It is neither in that it substends the "I-me" cognitive distinction and is that which gives rise to it. It is both in that this lived body functions both as an "I" and as a "me."[38] There is, however, no ambiguity if one remembers that the "I" as functional is the function of the creativity of the lived body. The "I" and the "me" emerge as two dimensions of the concrete subject that, in its vital functioning as creativity passing into fundedness or sedimentation, provides the ontological grounding for the sense of the dialogical stretch. The "I" and the "me" represent two distinguishable phases of the activity of the lived body that I am.

Turgendhat, noting a problem with Mead's position, has pointed out that the exercise of roles cannot handle the question, "Who do I want to be?" not in the sense of "a good teacher," or a "good parent," but in the sense of "What kind of a person do I want to be?"[39] Habermas attempts to capture this sense of concreteness that goes beyond particular roles through the distinction between role identity and ego identity, holding that this lies in the distinction between past and future: what kind of a person I have become as opposed to what kind of person I want to be.[40] These accounts may at first appear to involve merely a somewhat more restricted sense of role taking than is intended by Mead, for it may be held that the kind of person I want to be is handled by the kind of integration of roles I want to achieve.

The very sense of the question as to "the kind of person I want to be," or "the kind of integration of roles I want to achieve," however, requires not only more than the exercise of a series of specific roles but also more than any integration of roles. The sense of what Habermas calls "ego identity" as opposed to role identity seems to lie not in a distinction between the past and the future, but in the sense of the decentered self as the source of the role taking, the ontological "thickness" or living reality that takes, integrates, and changes roles but that can never be exhausted through this expression. It is the sense of the decentered concrete subject that operates in a present funded with the past and oriented toward the future.

Until now the discussion of the lived body, the decentered subject, has focused mainly on the dimension of creativity. Now the focus will move from this to the funded dimension of the lived body or decentered subject that gives rise to the "I-me" dialogue of reflective awareness. Mead holds that because the "me" is an object of reflection, in unreflective moments, the "me" disappears completely, while in some situations it is "relocated."[41] Critics of Mead have interpreted this, not without just textual cause, as indicating claims concerning not merely the self as the object of reflective awareness but the very existence of the self. Thus, in the first case, there is no self at all.[42]

Mead's own examples, however, can be used constructively in the present expansion of his understanding of selfhood. As he illustrates the relocation of the self:

> Tolstoi as an officer in the war gives an account of having pictures of his past experience in the midst of his most intense action. There are also the pictures that flash into a peron's mind when he is drowning. In such instances there is a contrast between an experience that is absolutely wound up in outside activity in which the self as an object does not enter, and an activity of memory and imagination in which the self is the principle object.[43]

Here again, the pragmatic understanding of temporality becomes relevant. As will be explored more fully in the final two chapters, for Mead a past is a past only in relation to a present. If we could recapture the past as it was, it would be a present that once was, not a past. Similarly, memory of the past is always from the perspective of the present, it is not a reliving of the past. The memories of a past self are memories as essentially related to, and emanating from, a present self. In remembering days of one's youth, for example, one sees them not as they were undergone at the time, but from the perspective of what one has become since.[44] These are images of my past because they are viewed from the perspective of a present self.

These situations indeed do not involve a present self that is a thematic object of reflection, nor do they involve a self actively but nonthematically engaged in dialogue in a present dialogical stretch. They involve, however, a present nonreflective sense of self, of the decentered subject as that to which my past is inherently related as my past, and which, at any moment, can give rise to the cognitive "I-me" dynamics relative to the present situation through the disruption of ongoing activity. What is of significance to note here is that Mead's examples of highly intense situations are not of those requiring thoughtfulness or deliberation but of those requiring just the opposite, the kinds of situations that one might say, in common parlance, require that one "act rather than think."

Not surprisingly, then, Mead also gives examples of situations involving either no awareness of self as an object[45] or a "relocation" of self because the activity is highly habitualized, as in the case of driving a car along an often traveled route. But although highly habitualized behavior, as not requiring deliberation, has much in common with the above examples of highly intense behavior, there is a special problem involving extremely habitual activity. In this type of situation it may appear that the ongoing adjustment to novelty inherent in sociality does not take place, even nonreflectively. Hence the sensed flow of the temporal dynamics involved in reconstructive activity, which at any moment can give rise to the emergence of the "I-me" cognitive dialogical relation, seems to have disappeared, for apparently there is no such activity.

Yet there can be no present without novelty and adjustment, the constituents of that sociality that characterizes any present in its passage. Thus, there can be no situation that does not involve some novelty emerging from a funded past and passing into new sedimentations, or, in other terms, some novel response of the "I" emerging from an old "me" and giving rise to a new "me," even though the adjustment be minimal. Mead recognizes this in his claim that "The resulting action is *always* a little different from anything which he could anticipate. This is true even when he is simply carrying out the process of walking. . . . That movement into the future is the step, so to speak, of the ego, of the 'I.' It is something that is not given in the 'me.'"[46]

A situation in need of no ongoing adjustment whatsoever could not be truly temporal. Every experience, even the habitual, is tinged with novelty and a sense of accompanying adjustment due to the very nature of time. In highly habitual behavior, the novel response of the lived body brings about an unreflective adjustment in terms of the fundedness that enters into the activity of the concrete subject. There is a non-reflective sense of the lived dimensions that underlie a possible "I-me" cognitive relation and that can bring about reflection at any time, reflection most likely to occur if a problem emerges that cannot be handled by highly habitualized behavior.

The concrete subject is constituted by structures of dispositional tendencies. These tendencies can be viewed as habitual in terms of the past, but are creatively anticipatory when viewed in terms of the future. Tendencies become habitualized in the sedimentations of the self, and such habitualized, organized attitudes or habits of response provide the structure that limits, but does not determine, the novel response that can emerge. As developed earlier, at times the emerging novelty and the resulting adjustment in the dynamics of the self are so rooted in past organizations of attitudes, even if originally structured through different types of activities, that reflection need not occur. The nonreflective awareness could be called either prereflective or postreflective. It involves a prereflective organization of activities in that it is that which is not, but can be, brought to reflective awareness. Yet this awareness is postreflective in that the very biological organization that is presently operating either in an extremely habitual or extremely intense nonreflective mode is what it is because of the structural organization growing out of activities of previous reflections, a structural organization that allows the bipolar dynamics of adjustment in the above cases to be carried out without reflective awareness.[47]

To have a self is to have an *ability* for reflective awareness and to incorporate dialogical activity involving the generalized other, and this this ability, once it has developed, is always "there" in the very structure of the lived body.[48] Once selfhood has developed, this ability or possibility is always there, and in the situations illustrated above, is experienced in the present as the nonreflective awareness of the temporally constituted dynamics of adjustment of the lived body, a nonreflective awareness that can give rise at any moment to the cognitive "I-me" dialogical relation when required, and that grounds the sense of the temporal dialogical stretch.

What needs stressing here is that the nonreflective experience prior to the development of selfhood is very different from the nonreflective experience that occurs after selfhood has arisen, for the latter involves an organization of ongoing adjustments structured by past "I-me" dynamics, a concrete decentered subject embodying the fundedness of past dialogical activities. Without the dialogical reflective level the social organization that enters into the sense of the concrete decentered subject capable of self-reflection could not come to be.

Rather, the sense of self that fringes experience is possible only through the taking of roles and the incorporation of the generalized other within the behavioral structure of the lived body. The development of selfhood is essentially dialogical, and any nonreflective awareness of self must presuppose the cognitive structure involved in taking the role of the other. But the very structure of the self as cognitive must be rooted in the temporal features of the lived body. The "I-me," subject-predicate, cognitive, reflective, or dialogical relation is rooted in the dimensions of creativity and fundedness of the lived body. The unity of the decentered corporeal subject in its dimensions of creativity and sedimentation is the unity of temporal experience rooted ultimately in the sense of the directionality of temporal passage. Yet such a unity can become the unity of *my* experience only through the cognitive activity involved in taking the role of the other. Time consciousness, role taking, and self-awareness are inseparably linked. The sense of the unity and directionality of the concrete self is founded ultimately in the sense of the unity and directionality of time.

At no time, then, is there a self directly accessible to cognition or reflection, nor is there at any time a psychical awareness of contents paralleling ongoing activity, each of which pragmatism is determined to avoid. Yet once the body that I am is given its due role, neither is there at any time the utter complete absence of self-awareness much less, as Mead at times extrememely claims, the absence of self. The non reflective sense of self as temporally structured decentered bodily subject does not parallel experience of objects but fringes it, pervades its very tone and texture, making my experience precisely *my* experience, my memories precisely *my* memories. In the nonthematic dialogical consciousness of the temporal present, the temporally founded sense of deliberative activity is again not some awareness paralleling what is taking place in the world, though neither is it a reflection on what one was or thought or did in a moment past, but rather it is a sense of the way one experiences a world in the temporal present.

This position does not deny the outward directedness of consciousness, nor is the awareness of the temporal vitality of the lived body something opposed to the perceived world but rather situated within it, partially constitutive of its character, and oriented toward it. The sense of one's existence in the world is the sense of the temporal dialogical stretch, whether the dialogue is fully operative at the cognitive level or incipient in the dynamic interplay of the creative and funded dimensions of the lived body. This sense of one's existence in the world is a sense of the temporally constituted decentered subject, which is at once a way of knowing and a way of being. The foundational constitution of the self begins, like the constitution of any meaningful awareness, with the primitive sense of temporal flow and develops into the commonsense grasp of oneself in terms of types of possibilities, potentialities, dispositional tendencies, or habits of response. Underlying any reflective grasp of this thick center of activity is a sense of its temporal dynamics in the durational spread of the present; the sense of the self functioning in its concrete creativity in the flow of time.

Such an understanding of the "thick" ontological dimension of the person as a dense system of possibilities shows also that our experience of other ontologically "thick" persons, whose commonness and uniqueness are manifest in their creative interactions, is a primary fact of experience. We do not begin with subjective contents as constitutive of our own person, and then from the outward behavior of another being "infer" that this, "like me," is a person having "subjective contents" or "inner intents and motives." The ontological depth and density of other persons with whom I interact is something I experience; I do not think or infer to them, but rather I interact with them. The unique ontological activity of the human organism that is not my own intrudes itself within my horizonal dimensions as its own kind of unique presence.

The above discussion provides a view of the self as inherently intersubjective or social, incorporating the role of the other in its very structure. The self is social not just because it is inseparably related to others, but because the very internal dynamics of selfhood evince the temporal dynamics of sociality through the way in which the I and the me, creativity and conformity, mutually affect and accommodate each other in an ongoing process of adjustment in the passing present. It also develops a view of the self that provides a new understanding of subjectivity, a contemporay understanding of a decentered subject that undercuts the problematics of the various versions of an intuited "I," of a transcendental ego, or of psychical contents paralleling or replacing an objective reality.

Further, in clearly rooting the cognitive level of the self in the temporality of the lived body engrossed in its cosmically embedded activity, it offers an account the account of self, not only in its existential but also in its cognitive dimensions, that is at one with an understanding of the cosmic dimensions of temporality, for the sense of one's self is inextricably interwoven with the sense of one's existence in an intersubjective and ontologically "thick" world. Time both flows through me and envelops me.

The way in which Mead's analysis of the cognitive self loses sight of his insights concerning time is intertwined with, and in all probability the result of, the way in which his understanding of the cognitive or dialogical nature of the self gets strangely detached from any ontological grounding in the lived body rooted in time and manifesting the temporal features of the temporal universe in which it is embedded. In losing contact with his insights concerning temporality, Mead slips into an implicitly atomistic perspective in his understanding of the self and, in so doing, lets slip away the sense of the self creatively functioning in temporal passage.

The above deepening and expansion of Mead's understanding of the self in terms of its temporal dimensions hopefully both highlights the immense power and potential of the general view of time housed within a pragmatic process philosophy and underscores the need to remain constantly diligent in its articulation and application to avoid the all too easy backslide into the all too treacherous, often all too elusive, terrain of temporal discreteness and its related self-defeating alternatives and dilemmas that Derrida so cleverly exploits and upon which his own tools of deconstruction so fully depend. Only if all vestiges of discreteness—be it knife-edged moments or atomic durational units—are rejected can personal identity through time be adequately grounded. And if the sense of personal identity through time is part and parcel of the temporality of the lived body in its ontological concreteness, then all vestiges of a subjective sense of time set over against time's objective reality is readily expunged. The following chapter will explore the implications of this temporal self for rethinking the nature of community and its temporal dimensions.

CHAPTER 7

Temporality and Community

Community and pluralism are often held to be at odds with each other, with a choice to be made between group conformity or individualism, community obligations or individual liberty, and this dilemma in large part structures much contemporary debate. Indeed, MacIntyre thinks America may well be founded on incompatible moral and social ideals. On the one hand, a communitarian vision of a common "telos," and on the other hand an ideal of individualism and pluralism. Thus, he holds that "We inhabit a kind of polity whose moral order requires systematic incoherence in the form of public allegiance to mutually inconsistent sets of principles."[1] These tensions, which tend to dominate reflections on community, are rooted in incompatible understandings of the nature of the self and its relation to the communal order that it inhabits, understandings that are in turn inextricably intertwined with implicit temporal issues.

Before turning to a pragmatically based account of this interrelationship, the ensuing discussion will very briefly explore the positions of Rawls, MacIntyre, and Rorty to elicit certain relevant temporal features. As a sketchy caricature, it can be said that Rawls's position exemplifies the individual as the source of important community arrangements, MacIntyre's the individual as the product of community arrangements, and Rorty's the freeing of the issue from ontological entanglements and presumptions.

Rawls's position is rooted in the self-interest-driven principles of abstract justice formed by isolated, presocial individuals operating through a veil of ignorance as to their own position in society. It emphasizes the primacy of the individual, and the social features stem primarily from the aggregate decisions of individual selves stripped of any particular attributes. While his position, at least in *A Theory of Justice*,[2] can be seen to involve an atomism in which separate individuals are ontologically prior to their unity, in *Justice as Fairness*, he reinterprets his position in light of communitarian criticisms. Here he argues that the concept of "artificial agents" deliberating in the original position is a device that does not imply any particular substantive conception of the self,[3] that the artificial or abstract self does not involve a metaphysical conception of the person.[4] Yet even this modification does assume that a self, abstracted out from its concrete relations or roles, can be coherently thought of as a functioning self, and that such a "self" can be a decision-maker.

In *Political Liberalism*,[5] which some interpreters view as a "later Rawls" that nullifies criticism of the "early Rawls," he in fact maintains his previous standpoint, claiming that the original position of the decision-makers, with its veil of ignorance and nonhistorical nature, is the basis for his discussion throughout the book.[6] And his qualifications of the abstract self remain the same as those offered in his 1985 article cited above. Thus, in spite of some new twists to his position, the core criticisms of it remain. The self that decides in Rawls's position is a peculiarly atemporal self, isolated from its historical attributes, ends, and attachments. Moreover, the frame that emerges from the debate is a peculiarly atemporal, rationally constructed frame imposed from above upon the contingencies of real life existence and isolated from the historically changing conditions within and among types of identities.[7]

Although throughout *A Theory of Justice* Rawls speaks of the formation of these principles though "our" intuitions, in *Justice as Fairness* he again modifies these claims in light of certain objections. In this latter work he clarifies that there is a certain ideal implied—that of Western liberal democracies,[8] and he allows that the basic values of the agent, now called "citizen," are not derived from basic intuitions but from an overlapping consensus. But Western liberal democracies seem to embody a pluralism such that there is no considered judgment that "we" must, as Rawls claims, "look for a conception of justice that nullifies the accidents of natural endowment and the contingencies of social circumstance as counters in a quest for political and economic advantage."[9]

Thus, while the formation of this frame is done from a basis of atomic self-interested individualism, the result is peculiarly nonpluralistic. What one has in fact is a rational, decision-making, atemporal, atomic self structuring an atemporal frame in which pluralistic differences have been left behind for commonly agreed to principles. People are essentially rational beings who have the possibility of reasoning their way to universally acceptable moral standards. Social structure is in some sense postulated in abstract principles, and social reasoning is by and large the application of the rule to the particular case. This ahistorical nature of the self is negated in MacIntyre's communitarian approach.

In *After Virtue*, MacIntyre endeavors to reintroduce the classical Aristotelian concepts of character, happiness, and virtue. To avoid Nietzsche and the contemporary chaos of an arbitrarily chosen plurality and incommensurability of values, we must return to Aristotle and the idea of telos. This teleological understanding of the human being was destroyed by individualism, in which human good was seen as prior to and separate from social roles rather than internal to these roles. MacIntyre does not seek a return to Aristotelian teleology, but argues instead for the necessity of a general "narrative" view of action in a reformulated idea of character. What is involved is not the passing of judgment on some kind of act itself, but seeing any act, as Aristotle did, as an expression of the more fundamental dispositions of the agent's character. This character is now understood as an ongoing narrative that determines the meaning of specific actions,[10] and it is impossible to judge the intent of any act by considering it in isolation from its context, or the agent's life narrative.

MacIntyre turns to a function of societies in which individuals are subordinate to the social structure implicit in complex activities or "practices," and in which individuals should endeavor to acquire the virtues necessary to achieve the goods internal to practices.[11] He recognizes that some practices are evil, and so while holding that practices are subordinate to the traditions by which they are transmitted, he does oppose the conservative, ideological understanding of tradition. A tradition "in good order" is "partially constituted by an argument about the goods the pursuit of which gives to that tradition its particular point and purpose."[12] As he states, "The fact that the self has to find its moral identity in and through its membership in communities . . . does not entail that the self has to accept the moral limitations of the particularity of those forms of community."[13]

Yet arguments about traditions and the limitations of community require some aspect of autonomy and some dimension of the self that outstrips social roles, a dimension of the self that remains unaccounted for in MacIntyre's position. It accommodates self-enclosed "traditions," each providing a context of rationality for working out its coherent community of civility. It does not provide for an autonomous self or creative agency that eludes reduction to roles and that engages in the ongoing reconstruction of the tradition. His communities are self-enclosed through their absorption of their pasts, rather than open to enlarging, self-transcending, future-oriented participation in a reconstructive process.

Rorty, in *Contingency, Irony and Solidarity*, argues that there is not and cannot be a reconciliation in the conflict between our private desire for self-creation and our public sense of moral duty, though this has been the aim of all "moral metaphysicians" from Plato on. He offers us an image of liberal society, an ironist utopia, that holds fast to its ideals while recognizing the incommensurability of values and its own historical contingency. In rejecting foundationalism, Rorty disallows a function for argument of principles and turns to a commitment to conversation that allows for both novelty and the inclusion of others.

But how does this openness take place? He holds that "All that can be done to explicate 'truth', 'knowledge', 'morality', 'virtue' is to refer us back to the concrete details of the culture in which these terms grew up and developed," and that "We have to take truth and virtue as whatever emerges from the conversation of Europe." He recognizes the validity of the objection that there is something very dangerous in the idea that truth is to be characterized as "the outcome of doing more of what we are doing now," for this "we" could well be the Orwellian state.[14] To answer this line of criticism, Rorty resorts to Habermas's claim that such a definition of truth works only for the outcome of *undistorted* conversation, and that the Orwellian state is the paradigm of distortion. But this of course will not quite do, for as Rorty well notes, Habermas offers transcendental principles, while he himself "must remain ethnocentric and offer examples." Rorty can only say: "'undistorted' means employing *our* criteria of relevance," where "we" are the ones who view truth and justice from the direction "marked by the successive stages of European thought." "We" all share this contingent starting point for our "ungrounded conversations."[15]

The undistorted conversation that is supposed to embody novelty, openness, and inclusiveness, then, turns out to involve an ethnocentrism that encloses us within our past. This enclosure is strengthened by Rorty's view that the self is "a tissue of contingent relations, a web which stretches backward and forward through past and future time,"[16] and human life is a "reweaving of such a web."[17] For, if the self is just the web, there is no creative agency to reweave, and its present cannot redirect the course of the future. The self and its conversation are alike ultimately enclosed within its past.

Rorty emphasizes the importance of increasing our sensitivity to unfamiliar sorts of people to prevent marginalizing them, to promote solidarity.[18] Yet the question remains as to how one becomes sensitive to "unfamiliar sorts of people" if one is stifled by ethnocentrism. His focus on imagination as the vehicle of increased sensitivity to unfamiliar sorts of people does not eliminate the problem but highlights it. He speaks of the meaningless, emotive function of imagination, as opposed to literal or cognitive content with its truth-value property.[19] This is in a sense a new twist to the positivistic dichotomy between the cognitive or meaningful and the emotive or meaningless, a twist that now views the dichotomy from the temporal perspective of the structurally located fixity of the old versus the utter capriciousness of a disconnected new.[20] Rorty's break between the meaningful and the meaningless, between the cognitive and the emotive, the literal and the metaphorical, is ultimately a break between past, present, and future.

The ability to grasp different contexts and enter into the standpoint of the other requires a creative agent that transcends the confines of ethnocentric enclosure in favor of perspectival openness. It requires also a temporal context that recognizes the equal importance, as well as the interpenetration of, past, present, and future or, in other terms, the openness of temporal perspectives, for imagination involves a link between what is and what can be brought forth in an oncoming future. In spite of, or perhaps better said, because of, the historicism of Rorty's position, we are left with a present that displays a disturbing atemporal quality. And, in spite of our supposed liberation from the illusion of any embeddedness of ourselves within a reality with which we meaningfully interact and which guides and constrains our ongoing inquiries about it, we are confined within a present that manifests itself as disturbingly ineffectual and unliberated. The past as it comes to the present is rich with possibilities for its ongoing reconstruction, but we are left with no access to a perspectival openness onto the richness of these possibilities and no creative agent who can utilize an openness onto possibilities in an ever more inclusive drive toward a novel future.

The discussion so far has attempted to show that each of the above positions yields, in its own way, a self that seems unable to execute its anointed community task, and this because it has become in some sense a peculiarly atemporal, ineffectual self within a peculiarly atemporal, ineffectual present. For Rawls, the self is locked in a present in which it functions in isolation from its past to make decisions that ignore the contingencies of the future. And, in spite of the different perspectives that color their understandings of the formation of the self by closed or ethnocentric communities, MacIntyre and Rorty alike offer an ineffectual self shut within an effectual present. Thus, for MacIntyre and Rorty, no more than for Rawls, is there a truly social self, for sociality involves a reciprocity of relations in which what is affected also affects; what affects is also affected. From the present pragmatic perspective it can be said—both in light of the earlier development of the self and in anticipation of what will be further developed in the remaining chapters—that in this sense sociality constitutes the nature of selves, the nature of the ontologically dense universe within which they are embedded and of which they are a part and, ultimately, the nature of the temporality that permeates them all.

Sociality, it has been seen, is the capacity for being in two systems at once, with the adjustments of each that this brings about; sociality itself is a process, the process of adjustment between systems that mutually affect each other. The self in its bipolar dynamics is constituted by an ongoing process of mutual or reciprocal adjustments of its two poles, each to the other. Its sociality consists not just in the fact that the self emerges in its very being from the context of others but also from the fact that there is, in the very structure of the self, a dynamic reciprocity of continuity and novelty, conformity and creativity. From this perspective, notions of an atomic self independent of its tradition and roles or a conforming self exhausted by its tradition and roles do not merely pose problems for resolving traditional tensions but represent ontological contradictions as well.

As noted earlier, Mead refers to the community or society that partially constitutes the possibilities for developing selfhood as the "generalized other." The following development will break with the conflation of community, society, and generalized other and will instead make a distinction between the generalized other, the social nature of experience as one of ongoing adjustment, and pragmatic community. True community requires not just myself and the generalized other, or just myself as partially constituted by the generalized other, but myself as both partially constituted by and partially constitutive of the generalized other. Further, it requires this in the dynamic and changing manner of ongoing socializing adjustment. The social nature of temporality permeates the dynamics of community.

Because of the very nature of the self, the individual is neither an isolatable discrete element in, nor an atomic building block of, a community. Rather, the individual represents the creative pole within community. The unique individual has been seen to both reflect and react to the common perspective in its own peculiar manner. In the continual interplay of adjustment of attitudes, aspirations, and factual perceptions between the common perspective as the condition for the novel emergent individual perspective and this individual perspective as it conditions the common perspective the dynamic of community is to be found. Thus the creativity of the individual can be contrasted with the conformity represented by the common perspective, but not with community. That which both founds and is founded upon the activity of ongoing adjustment is a community, and in its historical rootedness it develops its own particular organs for the control of the process. The ability to provide the means of mediating within the ongoing dynamics of adjustment between its two poles[21] constitutes a community of any type as a community.[22]

Novelty within society is initiated by individuals, but such initiation can occur only because individuals are continuous with others and with the historically situated social institutions of which they are a part. Part of the very life process is the ongoing adjustment between the old and the new, the stability of conformity and the novelty of creativity. The interrelation of novelty and continuity, creativity and conformity, stressed in virtually all the classical pragmatists, provides the conceptual tools for understanding how the uniqueness of the individual and the norms and standards of the generalized other are two interrelated factors in an ongoing exchange, neither of which can exist apart from the other. Because of the inseparable interaction of these two poles, goals for "the whole" cannot be pursued by ignoring consequences for individuals affected, nor can individual goals be adequately pursued apart from the vision of the functioning of the whole.

The development of the ability both to create and to respond constructively to the creation of novel perspectives, as well as to incorporate the perspective of the other, not as something totally alien but as something sympathetically understood, is at once growth of the self. Thus to contribute to the growth of community is at once to contribute to the growth of the selves involved in the ongoing dynamics of adjustment. Problematic situation can be viewed through the use of social intelligence in ways that enlarge and reintegrate the situations and the selves involved. In this way, a community controls its own evolution. Any authentic community involves a shared value or goal and, with this understanding of community, the shared ultimate "goal" involving the working character of universalizing ideals is self-directed growth or development, not final completion; a self-controlled evolving process, not a self-imposed fixed content.

Neither community nor the working ideal of universality can imply that differences should be eliminated or melted down, for these differences provide the necessary materials by which a society can continue to grow. As Dewey stresses, growth by its very nature involves the resolution of conflict.[23] Authentic reconstruction in cases of incompatibility must be based on the problem situation and the history within which it has emerged. However such reconstruction cannot be imposed from on high by eliciting the standards or traditions of a cultural past that may not contain the organs of resolution, but must be developed by calling upon a deepened attunement to a more fundamental level of human rapport.

While experience arises from specific, concrete contexts shaped by a particular tradition, this is not mere inculcation, for the deepening process offers the openness for breaking through and evaluating one's own stance. The deepening process frees intelligence from the bonds of rigidities and rule applications, as well as from the self-enclosed inculcation of tradition, allowing us to grasp radically different contexts, to take the perspective of the other, to participate in dialogue with the other, to utilize liberated possibilities. The deepening of reason can liberate us from the confines of tradition and rules, but this liberation is also the liberation of real possibilities that these traditions and rules have partially hidden from view; it is not a liberation from the ontological richness of the past and the possibilities it offers the present, but a liberation from a constrictive access to them.

No community is constricted by closed horizons from penetrating to deeper and broader levels of community, for perspectives open ultimately onto the common processive reality in which humans are embedded, of which their developing selves are a part, and with which they must successfully interact—hopefully in ways that enrich their evolving growth processes. In this way, expansion in breadth is at once expansion in depth. Such a deepening does not negate the use of intelligent inquiry, but rather opens it up, frees it from the products of its past in terms of rigidities and abstractions, and focuses it on the dynamics of concrete human existence. Pragmatism does not destroy reason but brings it down to earth, so to speak. Rationality is not fundamentally abstract, discursive, and calculative but concrete and imaginative.

This position thus agrees with Rorty in holding that imagination is vital in becoming sensitive to the other. Imagination, however, is not something separate from or opposed to rationality, but is part and parcel of it. It does not separate us from reality but rather is the vehicle by which we cast diverse creative perspectives upon reality in order to understand it. Rationality as concrete and imaginative is not working downward from rules to their applications, but working upward from concrete experience and decision making toward guiding hypotheses. Reason is radically historical not because it is the inculcation of a past but because it acts in the present through the appropriation of a living tradition that it creatively orients in light of a projected future. But it can do this only if imagination is not capricious but rather seizes upon real possibilities that a dynamic past has embedded in the changing present. As developed earlier, these possibilities are not confined within "subjectivity," for the living subject is part and parcel of the cosmos at large. There is a lived-through grasp of felt temporality opening onto a processive universe, and in this interactional context potentialities and possibilities function.

The social nature of temporality and reality in general, which permeates the dynamics of community as the socializing adjustments between self and other, negates the concepts of definiteness, atomicity, separation, or self-enclosure, whether applied to selves, communities, nature, or time. Rather, the open perspectivalism involved in the bipolar dynamics of self and community involves an efficacious openness onto both past and future and also onto an ontologically thick processive universe as the ultimate context for human interaction. It negates both relativism and historicism in favor of an ontologically grounded temporalism in which ontologically rooted perspectives emerge within the context of a past that presents itself in the richness of the present and that is oriented toward a novel future in an ongoing process of socializing adjustment. Historical rootedness is at once ontological rootedness, and these open temporal-ontological dimensions enter into, and should be fully utilized by, any tradition as it creatively articulates, develops, and reconstructs its characteristic features in particular ways.

The remaining two chapters will explore, in a more broadly based and systematic fashion, a pragmatic understanding of the temporal features of our lived experience and the universe at large, thereby developing the general context that in fact set the contours within which all other chapters took their shape. These two chapters bring to light more fully the way the pragmatic understanding of temporality involves a radical reconstruction of traditional ways of understanding and interrelating the key issues of time—discreteness and continuity; fixity and indeterminacy; potentiality, actuality, and possibility; past, present, and future.

CHAPTER 8

Temporality and Interpretive Activity: Continuity, Indeterminacy, and Experiential Flow

Within pragmatism, descriptions of temporal experience and more speculative cosmic claims concerning temporality are seldom clearly and explicitly distinguished, and trains of thought frequently slip back and forth between one and the other virtually unnoticed. This is perhaps the result of the pragmatic answer to a crucial question posed by David Weissman: "Is temporality itself the object or content of experience, or is something else being described, misleadingly, as temporality?"[1] He answers this question by stating his preference for the second alternative, arguing that "Temporal properties are represented within experience by considerations not themselves the core of temporality. . . . The experience-of-time is, I mean to say, a construction that is not in any way a direct perception of time itself."[2]

The pragmatic answer, as is well evident by now, is to be found in the first alternative. Temporality is not something that could be perceived by itself because the properties of time are the properties of temporal things or, perhaps better said, temporal activities. The features of temporality are the features of a temporal universe and its pervasive ways of behaving, ways that permeate the universe in general and various emergent realities within the universe, including human activity and human awareness of the universe in which it is embedded. A pragmatic reconstruction of temporality, then, must be at once a pragmatic reconstruction of our awareness of and embeddedness within a concrete temporal universe.

Yet, while there is an inextricable interrelation within pragmatism between the temporal descriptions of human experience and the more speculative metaphysical claims concerning cosmic time, the two cannot be collapsed without serious misunderstanding. The tendency to so collapse the two is in large part the cause of the not infrequent perceptions that pragmatism is by and large opposed to speculative philosophy, is certainly not a viable voice in ongoing discussions of process metaphysics, and is basically idealistic and anthropomorphic in what metaphysical bent it does manifest. Thus, the distinct focal points that separate this chapter and the next seem called for, but it must be recognized that this distinction is somewhat artificial, is made with great reservation, and can in fact be made only roughly and imprecisely. With this cautionary note in mind, the discussion will now turn to the focus of the present chapter.

In chapter 1 it was noted that Hartshorne objected to Peirce's emphasis on continuity and his use of infinitesimals in understanding the present, holding that it would require that we go thorough an infinity of operations of remembering even in the smallest stretch of time. This problem, according to Hartshorne, stems from Peirce's inability to make an adequate distinction between actuality as discrete and possibility as continuous. Weiss likewise takes issue with the pragmatic understanding of the relation between actuality and possibility and as a result he, like many others, objects to what he considers to be the overfuturism of pragmatic philosophy, holding that it supposes that what we grasp in the present are "solely future consequences." Hence, "were pragmatism correct, we could never verify anything, for an act of verification necessitates a grasp of the nature of what is now occurring and a reference back to the hypothesis, not necessarily a reference forward to something still in the future."[3] For similar reasons, Weiss objects that pragmatism "cannot handle the problem of the reality of the past."[4]

This type of criticism is present, with some variation, in Neville's objection to the pragmatic understanding of time. Instead of pointing to the "overfuturism" of pragmatism, however, he takes issue with the "overpresentism" of the position. As he states, the limit of pragmatism comes in its theory of time, for pragmatism is a philosophy of the present that is too biased to sustain a serious theory of temporal reality. Moreover, the most unfortunate situation for pragmatism, for Neville as for Weiss, is its inability to give an account of past things, for while the future collapses into a mere conditional projection of the present, the past has its reality only as a possible object of interpretation and is, like the future, reduced to its role in possible interpretation, which is an activity in the present. Where Weiss sees the present as well as past reduced to future possibilities, Neville sees the future as well as the past reduced to present actuality. As Neville summarizes the problem as he see it, pragmatism cannot handle temporal reality because it reduces metaphysics to what follows the structure of interpretation.[5]

However, pragmatism's radical reconstruction of the relation between discreteness and continuity, actuality and possibility, in the rethinking of the temporal dimensions of past, present, and future eludes these criticisms. Since the above criticisms all refer ultimately to the issue of interpretation, the following development will begin with the pragmatic understanding of temporality and related issues as these pertain to the interpretive nature of experience, most specifically with Peirce's analysis of time and infinitesimals.

Peirce's understanding of time has been of some interest to a minority of Peirce scholars, and of virtually no interest to those outside this minuscule circle—a situation that holds concerning the pragmatists' approach to time in general.[6] His view of time is inseparably intertwined with his understanding of continuity, for Peirce views time as the model for all other continua, as "the continuum par excellence, through the spectacles of which we envisage every other continuum."[7] Thus, his understanding of continuity in various contexts provide insights into his understanding of time.

Peirce's synechism, "the tendency to regard continuity, in the sense in which I shall define it, as an idea of prime importance in philosophy,"[8] is a doctrine that holds of the universe at large as well as of conscious awareness.[9] For now, however, the focus of concern will be on the role of continuity and infinitesimals only in his understanding of the immediacy of conscious awareness in the present. Peirce's interest in continuity has been divided in four stages, the pre-Cantorian, the Cantorian, the Kantistic, and the post-Cantorian.[10] The present analysis will not be concerned with these specific developmental divisions but will focus in broad strokes on issues that led to his changing views and the import of the position that finally emerged. Further, while Peirce's concern with continuity is inseparably linked with his mathematical interest in infinitesimals, the latter will be relevant only as it applies to his understanding of time and will not be evaluated from any technical mathematical point of view.

Derrida's problem is perhaps encapsulated in Peirce's insistence that we must be directly aware of the past as it recedes. If there were no such process of recession, then there would be a gap between past and present, and awareness of the past, and hence ultimately of the present, would be impossible. Thus, Peirce claims that "The world of memory is the world of time,"[11] but "were the instants independently actual, as they are in the Time of the analysts, memory would be a perpetual miracle."[12] The existence of memory is itself, for Peirce, a proof for continuity, for "if it were not so, nobody could have any memory."[13] As he states of the past, "it can only be going infinitesimally past, less past than any assignable date."[14]

According to this view, we are not aware of the present in a finite interval of time. Rather, "the present is connected with the past by a series of real infinitesimal steps."[15] That these infinitesimal steps cannot in any way be understood in terms of an indefinite succession of discretes but rather as a continuous flow can be seen from Peirce's somewhat detailed clarification. In turning to this, however, it should first be noted that Peirce makes a distinction between an instant as a point of time, and a moment as an infinitesimal duration, though as will be seen shortly, Peirce's manner here of speaking of this distinction is highly misleading.

> Now upon this interval follows another, whose beginning is the middle of the former, and whose middle is the end of the former. Here we have an immediate perception of the temporal sequence of its beginning, middle, and end, or say of the second, third and fourth instants. From these two immediate perceptions, we gain a mediate, or inferential, perception of the relation of all four instants. This mediate perception is objectively, or as to the object represented, spread over the four instants; but subjectively, or as itself the subject of duration, it is completely embraced in the second moment. . . . If it is objected that, upon the theory proposed, we must have more than a mediate perception of the succession of the four instants, I grant it; for the sum of the two infinitesimal intervals is itself infinitesimal, so that it is immediately perceived. It is immediately perceived in the whole interval, but only mediately perceived in the last two-thirds of the interval.[16]

The above shows, first, that there is a thickness and directional movement to time that is directly experienced in the passing present. Second, mediate perception involves interpretation. What must here be noted is that Peirce goes to great pains to develop the view that even a point of time, an instant as opposed to a moment, is not a knife-edged time slice. Rather, points of time, for Peirce, represent intervals, each of which can be understood as incorporating a beginning, middle, and end. Hence Peirce says that what is present to the mind, at any ordinary instant, is all of what is present during the moment of the occurrence of the instant.[17] In this way "the present is half past and half to come."[18] In this way, also, "we are directly aware of the flow of time, or in other words" of change.[19]

The radical significance of Peirce's view can be seen by the way it negates Augustine's position and its influence on successive thought. Augustine held that the present can have no extension precisely because if it did it would have to contain a little of the past and a little of the future. Similarly, change cannot take place in the present because change requires moving from an earlier to a later stage, and the present cannot contain an earlier or a later.[20] This exclusion of past and future from the present dominates the respective traditions of all those discussed in part I.[21]

An ordinary instant for Peirce is neither an atomic instantaneous point nor an atomic coming to be of a tenseless duration, but a marking off of an infinitesimal duration of time as a unit. Peirce does allow for a certain temporal independence to the instant, but he says that "It is not an utter, complete independence, not total independence"; rather, it is "an absolute independence *in certain respects*."[22] And the respects in which it is independent are functional. It represents the beginning of semiotic creativity, of spontaneity in the present, for "here new premises, not logically derived by Firsts, are introduced."[23] Further, it represents the beginning of novelty in general in the temporal flow of experience, though both of these "beginnings," in serving the function of providing for novel emergence, are in fact infinitesimal intervals that are inherently continuous. Thus, Peirce also emphasizes that "The striking of a new experience is never an instantaneous affair, but is an event, occupying time, and coming to pass by continuous process."[24]

The aspect of discreteness and continuity combined allow for the emergence of the present and, with it, the possibility of the emergence of semiotic and temporal creativity; but this is the case only because of the way in which continuity allows for traces of the past as possibilities for present creativity, possibilities that are "there" in the present and stretch into the future. Peirce's whole concern with infinitesimals and continuity leads him to the notion of the reality of unactualized possibilities that allow for a marking off of functional discreteness within what is not itself discrete.[25] As he summarizes, a continuum, "infinitesimal in duration, but still embracing innumerable parts, and also, though infinitesimal, entirely unlimited, is immediately present." Moreover, its unboundedness allows a vague possibility of more than is present to be directly felt.[26] As William James notes, in stating his agreement with Peirce concerning the basis for creative novelty, the mathematical notion of an infinitesimal contains, "in truth, the whole paradox of the same and yet the nascent other, of an identity that won't keep except so far as it keeps failing."[27]

Much is made by Whiteheadian scholars over the fact that James speaks of drops or buds of experience and in one of his later works holds that he will treat processes of change "no longer as being continuous, but as taking place by finite not infinitesimal steps, like the successive drops by which a cask of water is filled, when whole drops fall in to it all at once or nothing."[28] But if one reads this passage both in light of the text in which it is embedded and in conjunction with his lengthy footnote,[29] his supposed turn away from continuity and infinitesimals to finite drops is perhaps not the move to discreteness that it may at first appear, but rather a focus on the commonsense perception of change as a prelude to returning, at an explanatory level, to an account of continuity which incorporates infinitesimals.

As he states the problem, we are given a choice between "the opaquely given data of perception"[30] or the intellectualist absurdity of the unreality of change resulting from the mathematical division of a "point-continuum" into an infinite number of cuts. For "The mathematical 'continuum,' so called, becomes thus an absolute discontinuum in any physical or experiential sense,"[31] an experiential absurdity mathematicians have accomplished "by strictly banishing all infinitesimals."[3] James turns to "the buds or steps of process" that opaque perception "accepts as primal gifts of being," illustrating that they "correspond logically to the 'infinitesimals' (minutest quanta of motion, change, or what not) of which the latest mathematics is supposed to have got rid."[33]

Apparently James is here using drops or buds to show that, analogously to these, the continuum cannot be infinitely broken into a discontinuity of discretes but rather must be understood in terms of infinitesimal durations. Thus, the focus on experiential drops of change at the level of commonsense experience appears geared to supporting rather than denying the need to understand continuity in terms of infinitesimals if continuity is not to lead to paradoxes. This interpretation gains support from James's concluding statements.[34] He there observes that the question as to whether or not reality grows by abrupt increments of novelty is a "form in which the contrast between discontinuity and continuity confronts us," and his "business hereafter" must be to give a "perceptual account."

In light of his immediately preceding focus on the important role of infinitesimals for an adequate understanding of continuity and the analogous role played by drops of change in opaque perceptual experience, it is reasonable to expect that his final understanding of the relation of continuity and discreteness in a forthcoming technical account of our opaque perceptual awareness will incorporate an understanding of continuity based on infinitesimals in a way that does not negate but rather accounts for the experiential sense of novelty and change. As Peter Hare encapsulates the situation at this point, James is left with "the problem of how to make the discreteness of the growing infinite square with the continuity of the perceptual flux, and he ends the discussion with a promise (regretably a promise he did not live to fulfill) to work this out."[35] There seems no reason to think that this promise, if fulfilled, would not have incorporated a return to the understanding of continuity in terms of infinitesimals.

The whole conception of sequences of individual temporal units is an abstraction, for the movement from one interval to another is not a movement over discrete units. The continuity of time is not constituted by individual instants or discretes temporal units of any sort. This leads to Peirce's claim that Kant "squarely hit the nail on the head when he said that every part of a lapse of time was a lapse of time," but he failed to recognize that what he was saying involved not merely that "time has no parts that are not themselves times," but rather that "time has no ultimate parts."[36] For Peirce, "A true continuum is something whose possibilities of determination no multitude of individuals can exhaust," for it is not composed of individual units. As he observes, "Consciousness must essentially cover an interval of time," and "we are immediately conscious through an infinitesimal interval of time"[37] in which immediate consciousness is continuous. Thus Hartshorne's critical questioning of Peirce concerning "how many" acts of memory must occur in an infinitesimal stretch of time is entirely misplaced.

Peirce's blackboard analogy can be helpful here in understanding temporal and semiotic "discreteness" as a functional dimension of continuity constituted by traces of past and future as real possibilities present now, though not actual. Peirce views the blackboard as a continuity that makes everything upon it continuous. If a chalk line is drawn on the blackboard, there is an aspect of discontinuity, spontaneity, discreteness to it, but there is also an element of continuity to it. As he questions, "There is a certain element of continuity in this line. Where did this continuity come from? It is nothing but the original continuity of the blackboard which makes everything upon it continuous."[38] Hence, what is a singularity or discreteness in the containing continuum is itself a continuity in relation to further "discretes" potentially in it. Extending this analogy beyond the use Peirce made of it, the discreteness dimension of a first chalk line has an aspect of continuity that contains possibilities present now in a second chalk line, cut somewhere across the containing one. Yet the second chalk line, as it emerges in the present, has eliminated forever certain possibilities that before were real possibilities. They are no longer pasts for a next emerging chalk line. Further, the present chalk line is "rich with the future" because of the real present possibilities of future cuts that reflect upon its character; indeed, its character can be said to be constituted by those present possibilities.

Neither the past nor the future, in their aspect of possibility, denote settled entities that have been or will be, for just as past and future enter into the total character of the passing present as present possibilities, so the novelty introduced by the functional discreteness of the present enters into the character of the past and the future, changing the possibilities inherent in them as real possibilities. Any interaction or transaction between possibilities occurs in a passing present that grows out of a past and emerges toward a future. The passing present carries both novelty and creativity in a way that reaches backward into the past and forward into the future, both in terms of temporal creativity and semiotic creativity. Moreover, it can do this because the present is rich with traces, rich with the thickness of an inexhaustible continuum of possibilities stretching from the past toward the future.

This relation between continuity and novelty as it relates to the intertwining of our sense of time and the interpretive process is developed by Mead in a related but different manner that brings to light different features of importance. These concern the way in which the present understanding of continuity and novelty involves its own critique of Bergson's understanding of time, and involves as well the role of human activity in our sense of time or time consciousness. The way in which the functioning of quasidiscreteness represents the beginning of novelty in the present, as well as its role in creating a changing past, can be seen by turning to the temporal dimensions of Mead's concepts of adjustment and sociality, briefly touched upon in the previous chapter.

Sociality is constituted by participation in two systems and the mutual adjustments this brings about. The temporal nature of the adjustments that constitute the present is incorporated in the temporal understanding of sociality as the stage of transition or phase of adjustment between the arising of the novel within the old and the reorganization that gives rise to the new. This process of adjustment between the old and the new is constitutive of the passing present.[39] Mead's understanding of our awareness of time involves his understanding of emergence or novelty within experience and the reorganization of experience to adjust to the emergent. This requires, for Mead, neither the Bergsonian pure duration, nor the move to discreteness. As Mead is careful to point out concerning the significance of time in relation to continuity, "as one present slips into the next there must be some break, not *of* continuity, but *within* continuity."[40] The continuity is the condition for the novelty, while the novelty reveals the continuity though the oncoming adjustment that accommodates the novel, rendering it continuous with what came before. This relation of novelty and continuity is built into the very fiber of temporal experience: "there is a tang of novelty in each moment of experience."[41]

What is involved here is neither a mere continuity nor a mere duration, but rather a situation in which the past both conditions and adjusts to what is taking place in the present, and in which what is taking place adjusts to the oncoming future.[42] In this way, each new present brings about a new past oriented toward a new future. In this way also, while past and future emanate from a present, each temporal dimension is in a sense spread through every other. Thus Mead takes issue with Bergson's understanding of time as merely a stacking up of presents. As Mead encapsulates his view of time in opposition to that of Bergson, "It passes into another present with the effects of the past in its textures, not with the burden of its events upon its back."[43] The human being is understood as a praxis-oriented process whose acts of adjustment give rise to the sense of time. Its present is understood in terms of the happening of the act, which incorporates past and future in its process of adjustment. As Mead states, in stressing the temporal nature of existence, "The unit of existence is the act, not the moment."[44]

Mead's understanding of the distinction between the specious present[45] and the wider functional present, both of which are defined in terms of the act, parallels to a certain extent Peirce's distinction between the moment and the instant and, like Peirce's distinction, involves interpretive activity. Mead holds that the functional boundaries of the present are those of its undertaking—of what we are doing.[46] Further, any functional present is contextually located within, and continuous with, a more encompassing functional present with larger activities, and thus "we do not tend to have a sense of isolated presents."[47] Any functional present is extended by incorporation of pasts and futures relevant to the activity that marks out a present. These margins or horizons are extended both spatially and temporally through ideation, or the process of reflective awareness. Our functional presents "have ideational margins of varying depth, and within these we are continually occupied in the testing and organizing process of thought."[48] The pasts and futures are crucial to the activity that is constitutive of a functional present and are drawn into our experience of the functional present in the form of memory and expectation.

Whatever the span of the functional present, it is always wider than the specious present, for the functional present is constituted in terms of the ongoing act, while the specious present emerges for awareness when the act is disrupted in some way. Mead states of the specious present that "while it is an actual duration and not a knife-edged present, its duration is not that of the completion of the act within which the object is there but rather represents the disruption of ongoing activity. This claim concerning the relation between the functional and specious present is reflected in his following characterization. In a certain sense one's present takes in an entire undertaking through the use of symbolic imagery, and "since the undertaking is a whole that stretches beyond the immediate specious presents, these slip into each other without any edges."[49] But, for example, a loud noise behind an individual may mark out a specious present. "Its lack of relevance to what is going on leaves it nothing but the moment in which the sound vibrated within our ears."[50] The specious present marks the "limits of immediate change" within experience. "The limits of this span are uncertain because it so connects with the coming experience that there is no break in the temporal continuity, and because the passing experience goes over into memory imagery so imperceptibly that with difficulty he draws the line between them."[51]

There is, then, a twofold significance to the specious present. First, it represents the reconstructive phase in the breakdown of the act rather than the act's process toward completion; it comes to the fore of consciousness in the moment of the loss of the objectivity that characterizes the ongoing act in the functional present. It is the experiential residue within semiotic breakdown. Secondly, the experience of the specious present involves the direct experience of change. And, Mead emphasizes that the specious present or the "immediacy of the now" is never lost, but rather this experience of the specious present provides the basis for diverse constructions of the past and projections of the future.[52] The direct sense of change that is given in the specious present is never lost, and it provides the starting point for the constitution of a functional present temporally structured through praxis. The immediacy of the now, with its direct sense of change, is absorbed into the functional present in the ongoing act, though it can come to the fore in the disruption of the act. In this way, "There is an unalterable temporal direction in what is taking place. . . . There is a certain temporal process in experience. What has taken place issues in what is taking place, and in this passage what has occurred determines spatio-temporally what is passing into the future."[53]

The change that occurs in the process of adjustment is directly experienced in the specious present and forms the starting point for the constitution, through praxis, of the structure of the functional present. Mead holds that we can go back directly a few moments in our experience, though beyond that we are dependent upon memory images. In this way, we can directly experience part of the process of the passage of events. Mead, then, does not claim that memory and anticipation create the sense of temporal flow, but rather build on it at each end. To get beyond the specious present to the constitution of the functional present involves both memory and ideation. Memory here is not reproductive but reconstructive. The function of memory imagery is the filling out of our present perceptions and involves reconstructive activity. The reconstruction of the past lies in a present oriented toward a future.

What can be elicited from the above two brief, complementary sketches of Peirce and Mead is that time consciousness is intertwined with ongoing activity, activity that is shot through with interpretive dimensions. It involves a direct experience of the present as spread out and as incorporating the sense of change and vague possibilities, such that the present can be neither fully determinate nor isolated from past and future. Because of this direct experience of temporal flow there is an experienced temporal direction in what is taking place, and what is taking place involves a process of adjustment. Time cannot be understood metaphorically as a flowing stream moving from past to present to future; rather what is past or future to the subject is structurally related to the present—though not less real than the present, and as each present comes to be time adjusts throughout.

Time is experienced as a whole but as an incomplete whole. This wholeness of time's movement cannot be accounted for by a linear, unidimensional view of time or by viewing time as a passing of discrete units placed head to head; temporal passage is not a linear stacking up of presents. Nor is there any need for some linking together of past, present, and future, because each one of these dimensions essentially includes the others. Lived temporality thus incorporates a temporally extended present within which experience opens onto past and future and to which past and future adjust. Such an understanding renders impossible not only a fully determinate future but also a fully determinate present or a fully determinate past, but it provides for a thickness and directional movement to time that is directly experienced in the temporal flow of the passing present.

The semiotic activity of humans creatively structures the future in light of the past and, perhaps somewhat metaphorically, the temporal present creatively "intends" the future in light of the past. Both by their very nature are thickened by fundedness and anticipation, are rich with the real though unactualized possibilities inherent in the continuity that pervades the present. The very nature of the present is to mean, the very nature of presence requires signs. This understanding of signs in presence is viable because while Peirce and Mead look at intervals of time in a broader and narrower sense, they clearly develop a view of the narrow sense as itself a temporal spread that cannot be reduced to a point instant, and a view of this spread as providing a functional discreteness within a continuous process of becoming other.

In this way they decisively undercut Derrida's pseudo-alternatives, either pure identity or pure alterity, and with it the related alternatives, either presence without signs or signs without presence, allowing the living present to be understood from a richer account that genuinely includes the vital functioning of traces of past and future within the passing present. Moreover, because any temporal spread is a marking off of a functional rather than an ontological discrete, they avoid the ongoing Whiteheadian problem of the need to weld together discretes, an endeavor that can yield at best a quasicontinuum, in favor of the "emergence within" continuity of quasidiscretes. While Whitehead complained that Bergson's continuity had neither unity nor multiplicity,[54] pragmatic continuity can best be understood as having both unity and multiplicity. The multiplicity is constituted not by absolute discretes but by functional discretes, quasi-discretes whose functionality is that of providing novelty. Novelty is not opposed to continuity, but rather is made possible by it.

LANGUAGE AND TEMPORALITY

The understanding of the present as providing a "functional discreteness" or novelty that is the source of semiotic creativity rooted in praxis lends its features to the life of language. The emergence of language is entertwined with human interactions and is itself a form of bodily activity. Language involves the mediation of possibilities of response, and this, in Mead's words, "opens up the field of indication and reference out of which language can arise, though the two processes undoubtedly work together."[55] The structure of the interactive world of meaningful activity gives rise to the possibility of language, and our fundamental ways of thinking are abstractions from social interaction. The encompassing field of interest of organism-environment interaction provides the fundamental constancy that roots the constancy of all experience; the constancy and commonalty of language is intertwined with this interactive context, and both arise together in experience within the interpretive, anticipatory orientation of praxis, all incorporating alike its shared temporal features.

Language, as temporally constituted, partakes of the temporal features of continuity and novelty. In this way, language is not merely continuous with what was but actively enters into the constitution of new facts, new situations. Language as creative of new situations emerges from the novel activity of individuals, which in turn gives rise to new social contexts. This creativity occurs in a present context rich with the fundedness of past creativities with which it is continuous. Any creativity of language emerges in a present rich with the past. We are born into a common language that we creatively appropriate.

With language, as with all human behavior, we "build out" in both directions, incorporating the past and creatively structuring the future, and in the process bringing about a new past with changed possibilities. Language creates new situations by creating new meanings that are embodied in new ways of acting, giving rise to new sharable experiences. And, though there is a great range in our use of language, it is part of a social process by which we mediate the social situation through the understanding of meanings.[56] Mead forcefully brings this home in the claim that "A person learns a new language, and as we say, gets a new soul." That person must appropriate the peculiar attitudes of those that make use of that language, and in this sense become a different individual.[57] Language is expressive of a manner or style of behaving in a significant world. To truly appropriate a language is to understand the language, and that involves, ultimately, to live the life of the language, to live the social process that it expresses.

Such a linkage of language to the broader context of life sketches the radical difference between the pragmatic understanding of language and that of Derrida. Grammatical constructions do not stand alone but originate within and reflect the shared meanings of communicative interactions. Signs do not refer only to other signs within the system but are grounded in an extralinguistic reality that serves as their origin and ultimate reference. Language is temporally constituted, and as such incorporates the temporal dimensions of fundedness and novelty by which it changes and grows. Language can be creative precisely because it is continuous with its own fundedness and with the lived world of social interaction of which it is a part and within which it emerges, both of which it in turn restructures through its ongoing productive, creative usage. Language is rooted in a prelinguistic opening onto the lived world of human praxis and, as such, is intrinsically related to the world, to the language user, and to their intertwining in the temporal flow. There should be no wonder then that Derrida, in breaking up temporal passage into a play of discrete moments, forcing the choice between signs or pure immediate presence, between dire alterity or absolute identity, loses not only the present but both speaking subject and semantic reference.

The above understanding of temporal experience, language, and the interpretive process in general shows that the structure of experienced passage and the structure of interpretation are indeed linked within pragmatism. Moreover, Neville is perceptive in his stronger claim that pragmatism links its understanding of cosmic time, its metaphysics of time, to the structure of interpretation. For, as will be seen in the following chapter, the temporal textures of experience are part and parcel of the temporal textures of the universe at large. However, it would perhaps be more appropriate to say that interpretation is understood from a particular view of temporality rather than that a particular view of the temporal emerges from the structure of interpretation.[58]

Even with such a reversal, a key aspect of this structure involves the indeterminacy of the past. And, while interpretation may change our understanding of the past, and while semiotic or interpretive structures may adjust throughout to the creativity of present interpretations, to carry this claim to the cosmic nature of time may seem to support the contention that the pragmatic understanding of time is inadequate because it reduces metaphysics to the structure of interpretation. To clarify these issues, the ensuing discussion will turn from a quasiphenomenological exploration of temporal awareness and the interpretive features it embodies to a more full-blown reconstruction of the nature of time as rooted in pragmatic thought.

CHAPTER 9

Temporality and Cosmic Activity: Continuity, Indeterminacy, and Processive Concreteness

The beginning of the previous chapter noted that the pragmatic tendency to slip back and forth between descriptions of human experience and speculative metaphysical claims has led to serious misunderstandings concerning both the direction and the intensity of the speculative metaphysical claims housed within the position.[1] This tendency, clearly operative in Mead's discussion of temporality, provides a useful point of departure for the present chapter. For it relates to the issue of interpretation, highlights the general ambiguity in pragmatism concerning the metaphysical scope of its claims, and points directly toward a pragmatic metaphysics of temporality and the process philosophy this involves.

The earlier examination of Mead's understanding of temporal awareness showed that its present is characterized by the act, which brings about a temporal adjustment of past and future. This adjustment process is a feature not just of the constitution of human temporal awareness but of the universe at large. As he states, "If emergence is a feature of reality, this phase of adjustment, which comes between the ordered universe before the emergent has arisen and that after it has come to terms with the newcomer, must be a feature also of realty."[2] Nature for Mead is radically temporal, radically historical, though his own discussions are frequently ambiguous concerning the scope of the temporal features displayed in human experience.[3]

Mead's ambivalence concerning the cosmic extension of certain temporal concepts is very much in evidence in his discussion of the past, which is central to the present understanding of temporality. In his central doctrine of the nature of the past, he at times seems concerned to show that not only our viewpoint or interpretation of the past has altered, but rather the past itself has changed, for the past "in itself" is not a past at all. If we could recapture the past in its total concreteness, simply as it occurred, it would not be a past for any present, but would be simply a recaptured present, a present that once was. Its relation to the present is the very ground of its pastness,[4] for "One present slipping into another does not connote what is meant by a past."[5] Yet this turns out, within Mead's focus, to be explicitly directed not toward the relation between past and future in an independent ontological reality, but toward the reality of the perspective of the interpreter who elicits the meaningful structure of the past from the perspective of the present.

In one of the few times Mead tries to clarify his position, he again seems to lend support to the limitation of his claims to the perspective of the interpreter. He asserts that "The long and short of it is that the past (*or the meaningful structure of the past*) is as hypothetical as the future."[6] This clarification is misleading in that it covers up the radically temporalist nature of the understanding of the past in its cosmic dimensions that lurks in his position, and this covering up is intensified when combined with his further claim that in the sense that certain events have occurred, the past is obdurate, finished, final. In this sense "'the past' in passage is irrecoverable as well as irrevocable. It is producing all the reality that there is."[7]

The above two statements by Mead may at first appear to clearly distinguish a fixed past as independent of human knowing from the meaningful structure of the past as changing in relation to present interpretations. This reference to the finality of the past, however, is in fact a reference only to one dimension of the past in its independence of human interpretation. For the past as a past to the present, even independently of our present interpretation of the past, is a different past because of its relation to the present. It is this cosmic dimension of a changing past through ongoing phases of adjustment, haltingly and ambiguously put forward in Mead's position, that the following discussion will accept head on and develop, for the pragmatic rejection of temporal discreteness requires a rejection as well of all vestiges of the fully determinate, including a fully determinate past.

Before turning to this development, however, it may be helpful to focus briefly on a similar ambiguity of intent to be found in the work of Brian Martine,[8] who presents forceful objections to the claim made by the process view of time—by which he intends Whiteheadian process—that the past is wholly determinate. In developing his own original and insightful position he argues for both past possibilities and future possibilities, holding that the larger context of indeterminacy is what makes determinacy intelligible. In this context, he points out a problem for his own view. That is, if the past can "no longer be distinguished from the future according to the straightforward distinction between determinacy and indeterminacy, the possibility of providing a logical foundation for the 'arrow of time' as concretely experienced seems to fall away."[9] This is, of course, a problem to which any view that holds to the indeterminacy of the past must respond. And as is so often the case in Mead's writings, Martine also begins with an ontological question and concludes with an epistemic answer. For he ties this issue to an understanding of the way past and future possibilities relate to our knowledge.

Martine is very much concerned with epistemic issues as well as ontological ones, and his development offers many highly perceptive claims concerning his distinction between past and future possibilities. But that an original ontological problem is given an epistemic solution may well result from his view that possibilities are limited to past and future. While he speaks of past possibilities and future possibilities there is no mention of present possibilities. This is because his approach, though critical of Whitehead's understanding of the past, is linked to a fundamentally Whiteheadian position concerning the present, thus denying to the present a temporal spread inclusive of past and future. Contrary to this, what will be argued below is that, independently of any epistemic activity, the indeterminacy of the past allows for the arrow of time in large part because of the role of present possibilities constitutive of the temporal spread of an indeterminate present.

Peirce's blackboard analogy, utilized in the previous chapter but redirected for the present focus, serves well here. As Peirce develops the analogy in relation to cosmic evolution: "Let the clean blackboard be a sort of diagram of the original vague potentiality. . . . I draw a chalk line on the board. This discontinuity is one of these brute acts by which alone the original vagueness could have made a step toward definiteness."[10]

The previous chapter pointed out that there is an element of continuity in the chalk line that comes from the original continuity of the blackboard within which it emerged. What is a singularity or discreteness in the containing continuum is itself a continuity in relation to further "discretes" potentially in it. Extending this analogy beyond Peirce's use, it was further noted that the discreteness dimension of a first chalk line has an aspect of continuity that contains possibilities present now in a second chalk line, cut somewhere across the containing one. This second chalk line, as it emerges in the present, has eliminated forever certain possibilities that before were real possibilities. They are no longer pasts for a next emerging chalk line. Moreover, the present chalk line is "rich with the future" because of the real present possibilities of future cuts that constitute its character. In its aspect of continuity, then, the "cut" of the present contains past and future because it contains the continuities or possibilities from which it has emerged and the continuities or possibilities that extend into the future. In its aspect of discreteness, the "cut" of the present is the interactional event that helps direct the ongoing course of continuities, changing the possibilities constitutive of them.

In keeping with this analogy, interaction as distinct from possibility is a mode of behavior of a concrete processive continuum. Further, it is the interaction, not that which interacts, which is individual or discrete. What is intended can best be characterized not as actuality but as actualizations, as interactions or transactions that are actualizing activities. But these actualizing activities or transactional events are not ontological discretes, for actualizing activities are the activities of interacting continuities or possibilities, activities that alter them in an ongoing creative process.

Before continuing this point, two interrelated types of concreteness should be stressed. First, process is concrete in the sense of being indefinitely rich. It is "malleable" to our interpretive purposes, not because it is lacking in relations, but because it is abundantly rich in interacting events emerging from and giving rise to an indefinite richness of possibilities. Second, and more relevant for the immediate problem at hand, it is concrete in the sense that it is, as James so well express, the "living, moving, active thickness of the real."[11] Processive concreteness is the ontologically dense universe that intrudes within experience. It is the thick ontological density of, in James's terms, "the causal dynamic relatedness of activity and history."[12] Moreover, when one gets at the center of such a thick activity, it can be seen that the laws of the excluded middle and noncontradiction do not hold, for, as James again well notes, "place yourself at the interior thickness of the doing, and you see that in its concreteness it has the potential for breaking into the most contradictory characterizations."[13] Such a "breaking into" is not to be understood as a breaking up into parts but an emergence into novel traits.

Thus, processive concreteness in its various modes of activity has what Peirce calls a generality that "eludes" the principles of noncontradiction and the excluded middle,[14] or as Lewis notes, a generality that "overflows" such lines of demarcation,[15] not because it is either amorphous or ephemeral, but because it has the indefinite richness of a dense, opaque, thick, changing process. Continuity can best be expressed as concrete dynamic tendencies operative within processive concreteness, dynamic tendencies toward actualizations of varying types under varying types of circumstances. What the concrete does not mean, however, is the fully determinate.[16] No slab of processive concreteness is fully determinate. There is nothing that is fully determinate, finished, final, or individual.

One may object that "the interactional or actualizing event" fits the above characterization. But events can be understood neither as individual nor as fully determinate. Individuals are considered to have characteristics and, in fact, to have determinate characteristics that conform to the rigor of the laws of noncontradiction and the excluded middle.

But the event, the interaction itself, has no characteristics other than "bruteness." Any positive characteristics belong rather to that which interacts, portions of a processive continuity. Nor can the actualizing event be equated with existence. Spatiotemporal reality exists, but spatiotemporal reality, as processive concreteness, includes possibilities as well as events. Here the objection may be raised that existence is the arena of natural resistance, what we encounter, that facticity which is there and which our knowledge must adequately incorporate. However, we encounter possibilities as well as actualities, and these former as well as the latter have a brute facticity and natural resistance—a resistance that our knowledge must incorporate if it is to be successful.

Thus what exists, what is spatiotemporally real, is processive concreteness in its full behavioral dynamics. Existence, then, cannot be identified with the interactional event because existence includes much more than events. Existence and spatiotemporal reality are co-extensive.[17] The act of actualization as a discrete or determinate individual is an idealized abstraction similar to the idealization of a unit of temporal discontinuity. What it would be at an instant, it cannot be, because of its radically temporal nature. An event does have a unique character, but only because, as Dewey indicates, one cannot "isolate an event from the history in which it belongs and in which it has its character."[18]

The ambiguities of the term 'actuality' tend to parallel the ambiguities of existence. However, in converse fashion, actuality in modified form can be used in the present context to characterize the event, but is not applicable to processive concreteness in general. It may appear a bit strange to be willing to say that processive concreteness is real; it exists, yet cannot be characterized as actual. But if processive concreteness is characterized as actual, then we are faced with the confusing claim that actuality includes possibility, when actuality, as a philosophic category, is intended precisely to distinguish an aspect of reality other than possibility.

Arguments may be made that to refuse to call concrete existence actual is to fly in the face of common sense rather than to provide a metaphysics that accounts for our commonsense experience. Yet actuality as a philosophic concept that contrasts with possibility and actuality as used in common sense seem subtly at odds with each other, and may well be the source of categorial confusion. For example, common sense can well speak of actual possibilities, or say that something is actually possible, meaning by this that these possibilities are connected with real conditions, with spatiotemporal conditions, or with present conditions. Actuality as something opposed to possibility, then, can be be used only as an analytic distinction in the present context, and cannot be applied to any "space-time slab" of reality in its concreteness.

Actualizing or transactional events, as indicated earlier, can best be characterized not as actuality but as actualizings or actualizations; they are the interactions or transactions among possibilities. Also, it must be stressed that the interactional or transactional event cannot be understood in terms of two self-contained things that come together, but rather it is an aspect of a contextual whole in which each ingredient, including the interactional event itself, is what it is because of the nature of the contextual whole. This category of "actualizing" leads to two diverse sets of potential problems.

First, it may be held that by changing "the actual" to the activity of "actualizing," concreteness seems to have disappeared. Actualizing activities actualize possibilities, while novel possibilities emerge and develop from actualizing activities and a "substratum" of possibilities. The concretely rich matrix of possibilities and actualizing activities seems to have become somewhat ephemeral. However, to exchange the temporal dynamics of an ontologically dense universe for an ephemeral matrix of abstract possibilities and actualizing events is to forget that these concepts represent modes or ways in which an ontologically dense processive concreteness functions or acts.[19]

The way in which the concrete does not mean the fully determinate has been clarified above. No slab of processive concreteness is fully determinate. Hence again, in the sense that actuality indicates the determinate, the finished, the final, processive concreteness cannot be characterized as actual. Nor can actualizing be understood as a process that culminates in actuality so characterized. The philosophic concept of actuality seems more and more to be one of those categories that emerged in the context of substance philosophy but leads to confusions within the context of a metaphysics of activity, for there is nothing that is fully determinate, finished, final, except the actualizing events, the events of actualization, and these are the events of interacting continuities and thus are idealizations from a history and process of becoming other.

This statement immediately brings in a second potential problem: the charge made frequently, and exemplified in the earlier criticism by Weiss, that pragmatism overemphasizes the future because of its focus on possibilities or potentialities. Here again is evidence of categorial confusions. There is a very great difference between "future actualizings," "future possibilities of actualizings," and "*present possibilities of* future actualizings," though these (and especially the latter two) are often used interchangeably. The difference between present and future is not the difference between the presence or absence of possibilities, for past, present, and future can all be viewed in their dimension of possibility.

A pragmatic process metaphysics of activity does not put undue emphasis on the future and does not require that future actualizations in some sense be real now, but rather requires that *present possibilities for* future actualization be real. Future as well as past are drawn into the present for pragmatism, but they are drawn into the present in the form of possibilities engaged in an ever-advancing process of actualization, all emergent from the past and projecting toward a novel future. Again, the concept of fixed, determinate actuality seems to muddy or, perhaps better said, stagnate the waters. But its anchorage is strong indeed, and accounts for the criticisms of Neville and Weiss concerning the pragmatic overemphasis on the present or the future respectively.

A further development of the interrelation of possibilities, and events could call upon Dewey's concept of situations or Mead's understanding of systems and the adjustments they involve. However, the ensuing discussion, in extrapolating from, will leave behind the language of, chalk lines, situations, or systems and, in the spirit of neutrality, introduce the term 'matrix' to briefly sketch an imaginative model for understanding an interrelation applicable, in broad strokes, to the general spirit of a pragmatic understanding of cosmic time as here developed.[20] The term 'matrix' as used here is not intended in any mathematical sense, but rather is meant to indicate any delimited cluster of possibilities and ongoing actualizations. These matrixes can be discussed from two perspectives, their cohesiveness and their diffusiveness. In this context a distinction will be made between possibilities and potentialities, the lack of which up to this point may have seemed a bit glaring.

Any matrix has a cohesiveness or integrative directional flow, yet in its indefinite richness its concrete possibilities are functionally diffused throughout and embedded within a network of other matrixes. The functional diffusions of any matrix permeate other matrixes, while conversely, any matrix is permeated by the functional diffusions of others. These interpenetrating functional diffusions shape the changing contextual or horizonal dimensions of a matrix.

The phrase 'diffused throughout' another matrix is used rather than the phrase 'part of' another matrix to avoid any metaphor that might suggest a building-block model. A matrix of more encompassing scope is not composed of parts. Rather, while any matrix can be understood in terms of both its integrative directional flow and its functional diffusions, any functional diffusion within the context of a matrix is itself a matrix that has an integrative directional flow and that is functionally diffused throughout other matrixes. Hence, it has a matrix as intricate as, but partially different from, the matrix of which it is itself a functional diffusion, although its continuance may be dependent upon its role as a functional diffusion. The distinction between a functional diffusion and an integrative flow, is thus contextually relative. There are no ontological building blocks. Rather, there are various scopes or levels of interpenetrating matrixes. Also, the term 'diffusion' is intended to avoid the concept of spatial containment. Any processive concreteness is diffused throughout its temporal relationships. Such temporal relationships involve pasts, presents, and futures.

Because of the functional diffusions within a matrix, the possibilities that are constitutive of its integrative character, and that can be termed its potentialities, are always operative within a network of broader possibilities. Conversely, what are potentialities within the context of one matrix are, as functionally diffused throughout another matrix, part of a contextual range of possibilities within which other potentialities constitutive of other matrixes operate. Thus, potentialities and possibilities are alike dynamic or dispositional tendencies, but dynamic tendencies and the transactional events that actualize them occur in partially intermeshing matrixes, and the difference between potentialities and possibilities is a function of the role of dynamic tendencies in relation to a particular matrix. The potentialities constitutive of a matrix are at once functionally diffused throughout a multiplicity of possibilities, each of which is a potentiality within a matrix that it helps to constitute.

The nature of matrixes as interfused or intermeshing continuities accounts for the ontological character of novelty or emergence, for the novel emergence is a function of interacting dynamic tendencies. And as the novel emerges these matrixes adjust throughout, bringing along in their temporal advance a new past oriented toward a new future. Moreover, they can adjust throughout because in their concrete spatiotemporal existence they are indeterminate throughout. From the pragmatic framework, the alternative to Bergson is not the ultimacy of discretes but the recognition that time moves as a whole and with depth, and that its movement is affected by the emergent present, which is both the effect of and affects its textures. The emergence of the novel arises from a past to which it does not belong, but in its arising it changes the nature of the past by creating new relations with it. While spatial metaphors are inadequate to deal with time, it is virtually impossible to deal with time without some spatial metaphors. But as opposed to saying of increasing temporal "distance" that the line of time grows longer, it may be more appropriate to say the layers of time "thicken" and adjust.[21]

The character of reality as relational and temporal constitutes at once its character of novelty. The event, as the idealized limiting point of the temporal present, is the idealized moment productive of interactive novelty. Possibility, as the "substratum" or contextual horizon for alternative realizations of the potentialities of cohesive matrixes, is the horizon for novel emergents. Potentiality, as the creativity of dynamic tendency in its "loose play," in its indeterminateness for the selectivity of alternative possibilities of actualization yet to be realized, is at once the foundation of lawful endurance and the bearer of a novel future. In its ongoing unfolding potentiality changes the range of real alternatives "there" for its further development. In this way, the intermeshings of matrixes provide for processes that progress toward new foundations of intelligibility, a kind of process in which the telos of the process itself changes in its ongoing development.

Moreover, any matrix is at once one and many. In its integrative flow it has a unity, but through its functionally diffused dynamic tendencies it is also in a continual process of becoming "other," and in its dynamic "loose play," in its "selectivity" from a context of possibilities, it both affects and is affected by the temporally emerging present as events that won't keep, because the isolated event is an abstraction from a history and future of becoming other. The past and future are already part and parcel of the present "least bit" of reality, just as they are part and parcel of the present "least bit" of experience, for there are no absolutely least bits of temporal flow, any more than there are least bits of reality or least bits of experience severed, through their discreteness or determinacy, from that which is other. As such, there is at once, to use Dewey's terms, the union of the "settled and the unsettled, the stable and the hazardous."[22]

Because of this, the unsettled, the unstable, the novel emergent is not a characteristic that holds only in relation to human noetic activity, but is there regardless of such a presence. The indeterminate situation is objectively indeterminate, independently of consciousness. As a result, it is potentially problematic for the conscious organic function whose horizonal dimensions it enters. Indeterminacy is ontologically pervasive. What has no ontological status is the fully determinate, the permanent, or the discrete. What are in a sense "determinate," "permanent," or "discrete" are the actualizing events, the events of actualization, but these are idealized abstractions from interacting continuities, not ontological ultimates. Reality is flowing and continuous and as such, indeterminate; in its indeterminacy it eludes the logic of identity.

The idea may be held that if all possibilities as independent of noetic activity are dynamic tendencies of processive concreteness, then the universe is in fact deterministic. Though possibilities bring novelty into any matrix, such novelty holds only from within the perspective of a matrix and disappears in the light of the overview of intermeshing matrixes. This, however, misses the character both of the intermeshing of matrixes and of true temporality. James's characterization of the universe well expresses the character of the intermeshings when he observes that the universe is "many everywhere and always, nothing real escapes from having an environment . . . the sundry parts of reality may be externally related."[23] The nature of functional diffusions is such that any matrix incorporates an environment, part of which need not be internally related to it. While every functional diffusion is itself a matrix of more limited scope, at least in some respect, than the matrix of which it is a functional diffusion, there is no one matrix that "contains" all the others as parts to a whole, for the nature of functional diffusions is such that intermeshings are always partial and incomplete.

Though there is no relation of whole to parts, there is indeed an interwoven continuity of intermeshings. But true novelty does not deny but rather requires such continuity. James remarks, in expressing his agreement with Peirce, that to "an observer standing outside of its generating causes, novelty can appear only as so much 'chance,' while to one who stands inside it is the expression of 'free creative activity.' . . . The common objection to admitting novelties is that by jumping abruptly, ex nihilo, they shatter the world's rational continuity."[24] But, as he continues, novelty "doesn't arrive by jumps and jolts, it leaks in insensibly, for adjacents in experience are always interfused, the smallest real datum being both a coming and a going."[25] And once again, James understands this position as tied to the notion of an infinitesimal as disallowing the fully discrete or determinate.[26]

The passing present brings novelty in its flow, a novelty that is not abrupt, but that in its emergence is part and parcel of the context of processes that preceded it and that follow from it through ongoing adjustments. If only past and future actualizings were involved, then the passing present would be merely a deterministic push from the past or pull from the future. Past and future, however, are conditions involved in the present not in their aspect of actualization but in their aspect of continuity or possibility.[27] Neither the past nor the future, in their aspect of possibility, denote settled entities that have been or will be, for just as past and future enter into the total character of the passing present as present possibilities, so the hereness and nowness of present actualizings enters into the character of the past and the future, changing the possibilities inherent in them as real possibilities. Any interaction or transaction between possibilities occurs in a passing present that grows out of a past and emerges toward a future.

The passing present carries novelty and creativity in a way that reaches backward into the past and forward into the future. In exhibiting both past and future, the passing present alters both, for past and future are continually reconstituted by their relations to what emerges within the present, and these relations change as what emerges changes. The degree of change between past and future is of course vastly different, for the past incorporates as well the dimension of the obdurateness and finality of transactional events that have in turn set contours for ongoing dynamic tendencies.

The potentialities and possibilities of a past present are gone, as are the actualizing events, but what must again be stresssed is that the possibilities and potentialities of what is past relate to the present as the conditioning factor of its pastness. The past, as the past to the emerging future, exists now as the potentialities and possibilities diffused within the present, and as any present to which it relates changes, so it changes, no matter how slight this change may be, for its potentialities and possibilities that make it what it is in its total concreteness change. As long as there is a present actualizing possibilities that can be actualized but need not be, there is a changing past, for with every emergent present certain possibilities that were in the past real possibilities are no longer possibilities for future emergent presents, while possibilities that were once not there come to be. Such changes are independent of our perspectives, though the changing past, as a past in relation to an emerging present, has like all matrixical relationships an indeterminateness born of an indefinite richness that the human mind must make intelligible through the interpretations that render indeterminate situations more precise.

It has been objected that this type of pragmatic view gives an ontological primacy to the present and rids the past of its integrity. The present does not have ontological priority over the past or the future, but is the "location" of the ongoing efficacy of past and future as well as of the novelty to which past and future adjust. There is no ontological primacy of present over past and future but a continuity among past, present, and future such that each is spread throughout the others, and each is what it is because of this mutual diffusion. In a sense it can be said that the position gives a functional priority to the present, but even this is misleading. For the present can function as it does only because its activity incorporates past and future. The past has its own integrity, but integrity and fixity are not synonymous.

Any contextual matrix, as a cohesive matrix embedded in a concrete multiplicity of possibilities, is both unique and uniquely unified. Yet the integrative directional flow of the cohesive matrix grounds a basis for similarity of function with other concrete cohesive matrixes. Hence in its concrete lawful mode of behavior, it grounds a basis for the conceptual organization of generalizing or universalizing laws. Laws express recurrent types of conditions under which integrative directional flows of particular types function and, in so doing, express the conditions for the emergence of qualitative events in the ongoing interactional process. Causal laws and qualitative emergents are part and parcel of the same natural processes.

While temporal ordering is dependent upon the reflective ordering of conscious organisms, temporal quality is pervasive of every matrix. It is this temporal quality in its ongoing flow which is experienced in the prereflective awareness of our praxis-oriented existence, and which is so easily dismembered by the reflective attempt to capture it. As Dewey stresses, temporal order "is a matter of relation, of definition, dating, placing, and describing," and as such is discovered through reflection, while temporal quality "is an immediate trait of every occurrence whether in or out of consciousness."[28] And emerging events are qualitative in other ways because of their fundamental temporal quality, for they carry with them their histories. Additionally, events are abstractions from processive continuity, and thus the temporal flow of interacting matrixes carries with it a continuity of changeable and changing qualities.

One of the qualities that emerge in the interactions of matrixes, one of which is organic, is the quality of value.[29] This quality, like all qualities, is inherently temporal. It carries with it the history of an organic cohesive matrix as well as the history of its intermeshing environments.[30] The very notion of value carries with it the temporal advance of activities toward consummatory experiences. The dynamic tendencies involved in the temporal stretch of an organic matrix are at once anticipations of future qualitative occurrences as consummatory experiences. The temporal stretch of human behavior not just as anticipatory but also as consummatory is experienced within the durational flow of felt temporality. In this way value is an emergent and pervasive feature within the context of intermeshing matrixes constitutive of organism-environment interaction, and can be located neither within the organism alone nor within the environment alone.

Further, even the most rudimentary valuings are neither isolated nor brute. For the valuings of anticipations satisfied are valuings emerging from satisfactions that have been partially shaped by others with which one's horizonal dimensions are intermeshed. Moreover, in the ongoing temporal flow and the anticipatory activity it incorporates, even a temporal spread incorporating the most rudimentary anticipation is tinged with the funded character of past fulfillments or frustrations. The temporal stretch of behavior carries the felt continuity of anticipation and fulfillment, as well as the "being with" of the other, into the very texture of valuing experience. Experience is through and through both temporal and social because reality is through and through both temporal and "social."[31]

The human organism, then, finds its ontological status within the reality of intermeshing matrixes. Organism-environment interaction is a specialized instance of the intermeshing matrixes that constitute processive concreteness in general. The human organism, as a matrix become conscious, has the unity of an integrative directional flow, yet is in a constant process of "becoming other" due to the functional diffusions of other matrixes, both conscious and unconscious, that contour its horizonal dimensions, and due conversely to its functional diffusions that are spatiotemporally spread throughout any matrix that enters its horizonal field. In this way it both is affected by and affects its environment. The organism and its environment are what they are in their respective characters because of these intermeshing contexts. Some of the matrixes that both affect and are affected by the character of the conscious organism are other conscious organisms. Like all matrixes, the human matrix is inherently social. In distinction from inanimate matrixes and less sophisticated organic matrixes, however, the human organism can consciously direct its creative appropriation of what enters its horizonal dimensions. The conscious and intelligent guidance of creative appropriation converts novelty into human freedom.

From the above backdrop the focus will turn more explicitly to the unification of the temporal features manifest in anticipatory interpretive activity with the temporal features manifest by processive concreteness, a unification seen by critics of pragmatism as a key source of its temporal inadequacy. As indicated earlier, the present pragmatic approach does indeed provide an interrelation between the structure of interpretation and the structure of reality. This relation, however, does not "reduce metaphysics" in any way,[32] but rather provides a metaphysical grounding for the structure of interpretation that operates within the conscious human matrix.

Previously it was noted that the structure of interpretation as a temporal process is inherently creative and anticipatory. The most rudimentary experience as temporally grounded displays these features and, in displaying them, the structure of experience follows the structure of scientific or experimental method. This experimentalism, as already shown,[33] involves a selective, creative organization of experience that is purposively transformative of its environment by rendering more determinate an indeterminate or problematic situation, and that is directive for ongoing activity. And its truth is tested by its workability or, in other terms, by the fulfillment of anticipations in the ongoing course of experience. This latter feature is the case precisely because its significance in the present lies in the present possibilities for future experiences that it contains.

The description of the temporal dynamics of intermeshing ontological matrixes in general shows that a dynamic tendency is "selectively and creatively anticipatory" of the oncoming present. It renders more determinate an indeterminate and hence "problematical" situation,[34] and in so doing is transformative of its environment; it is directive of the course of ongoing activity and, as being a present possibility for future emergent activities, its significance in the present points toward the future. It is "purposive" in that intermeshing matrixes develop and change through their own ongoing adjustments, adjustments that are the results of neither a fully determinate past nor a predetermined future. This is so because, as stated above, the flow of dynamic tendencies in its "loose play," in its indeterminateness for the selectivity of alternative possibilities of actualization yet to be realized, is at once the foundation of lawful endurance and the bearer of a novel future, functions that are mediated and rendered continuous through actualizing adjustments in the present. The passing present of cosmic activity thus "intends" or "means" the future in light of the past, and does so through an adjustment process by which it reorganizes the past through the new relations to the past that it creates in its own novel emergence.

In this way, a pragmatic process metaphysics of action provides an understanding of the way in which the temporal dynamics embedded in the spatiotemporal, concretely rich universe provide the metaphysical foundation for the temporally founded dynamics of conscious interpretive activity operative in concrete human existence[35] or, in other terms, for the temporally founded dynamics of the anticipatory nature of experience as experimental.[36] The temporal structure that belongs to the universe belongs as well to the interpretive processes that emerge within it. Human habits of response, which are for the pragmatist the living embodiment of meaning, are precisely "ontologically thick" dynamic tendencies structuring emerging activities in the context of alternative possibilities for ongoing actualization.

A disposition or habit as a regulative rule is something whose possibilities of determination no multitude of actually generated instances can exhaust. The primordial sense of the unactualized creative potentialities of habit as experienced in the actual temporal continuity of the durational present gives a concrete sense of a real lawfulness that governs unactualized possibilities, of potentialities that can never be exhausted by any number of actualities. Additionally, the sense of unactualized possibilities embedded in dispositions or habits brings a sense of real alternatives—the could do or could be otherwise—into the heart of perceptual awareness, providing a sense of freedom and what Lewis refers to as a "primordial sense of probable events."[37]

In the immediate flow of experience as an interactive unity of humans and that broader context within which they are embedded, these temporally founded traits cannot be uniquely attributed to one or the other. There is an interactional unity at the heart of temporal experience that contains a two-directional openness: the primordial sense of the pulsations of temporal existence opens in one direction toward the temporal pulsations of cosmic activity and in the other direction toward the structure of the temporal pulsations of the human way of being within whose purposive activity the temporal dynamics of the universe at large manifest themselves at an emergent level of activity. The sense of temporal flow embedded in interpretive activity is at once a sense of the temporal dynamics of a processive universe, with our own dispositional tendencies or "living meanings" as part and parcel of it. These dimensions cannot be understood from the pragmatic framework unless the search for the fully determinate, the fully fixed, the fully discrete yields to the basic pragmatic intuition of the continuity and indeterminacy that pervades a radically temporal universe in the fullness of its space-time concreteness.

In the primordial sense of directional flow, the primitive "feel" of real potentialities structuring the very character of emerging activities grounds the "if-then" relation of human activity to anticipated ends and provides the sense of real causal efficacy and the arrow of time. The if-then of the relation of human activity to anticipated ends is a relationship founded in the temporal dynamics of the universe and permeating our primordial sense of temporality and its directional flow.

The sense of the future in general implied by the structure of human behavior as anticipatory is neither a sophisticated nor primitive induction from past experience but is at the heart of experience in the durational present. Moreover, the experience of the durational present, as incorporating both past and future, founds our very sense of the past, and memory itself, as a continuity within which present experience emerges, colors the emerging experience in the passing present. Within a context of the ultimacy of discretes, not only is it the case that habit no more than cause can be experienced, but the memory of past cases needed to provide specific habits of expectation remains inexplicable, and hence these specific expectations themselves becomes inexplicable. Thus, the experience of the temporal dynamics operative in the durational present, dynamics that involve mutually interpenetrating adjustments of past as well as future, accounts for our memory of the past and our expectation of the future, and so for the very possibility of interpretive activity of any sort.

In this way, the sense of temporal passage grounds the causal explanations, abstract inferences, and experimental developments of physical science that, through its use of mathematical, spatialized time, leaves behind for its own valid purposes the very sense of temporal passage that grounds its endeavors. The time of nature within the world of science cannot be confounded with the temporal passage of the natural universe in its processive concreteness. To speak of the structure of temporality or a spatiotemporal slab of concrete reality is not to indicate the ordered content of some particular mathematical system, but rather to indicate the ongoing unfolding of concrete cosmic time that ultimately both grounds the possibility of, and renders "peculiar" to common sense, the abstract, reversible time of physics.

Nor can the character of the indeterminate generative activities and the causal relations sustained within temporal passage be conflated with the character of the causal relations holding among the events in the world of science. Causal connections are always expressed as relations among particular types of objects or events, and the nature of the events or objects being connected enters into the very understanding of the nature of the causal relationships sustained. Though the brute interactional aspect of nature can be abstracted out and creatively organized to yield the causal connections and space-time events of science, the concrete processes of nature involve temporally directed generative activities that elude or overflow such abstractions.

The concrete processes of nature are a dimension of our everyday natural environment; the temporality of the universe at large is an essential dimension of our "felt" or lived sense of temporal flow; our felt sense of time *is* cosmic time as it both pervades and engulfs us; lived temporality is part and parcel of a temporal universe. Nature as a system of temporally structured scientific events is a reflective, creative, abstract explanatory net cast upon a temporal universe, one that works well for the purposes of scientific explanation. If this explanatory net is substituted for the temporally grounded features of a processive concrete universe, or becomes in any way the model for understanding the constitution of "real" temporality, then the divide between time consciousness and cosmic time will remain impassable.

At this point it will be helpful to consolidate some brief claims scattered throughout the above concerning the relation of spatiotemporal concreteness to our modes of grasping it and rendering it precise. This will also involve saying something more about the relation between potentiality and possibility. Potentiality and possibility have been seen to be inherently temporal, but whether or not one takes them as temporal per se, they are intimately intertwined with temporal issues. Hence Charles Sherover suggests that "on one level at least, the old ambiguity between potentiality and possibility may well rest on a lack of discernment of temporal modes."[38] According to the present line of development, however, the distinction between potentiality and possibility is not based on a distinction between temporal modes. Rather the distinction is a contextual one based on the interrelation of dynamic tendencies within a developing matrix and on the "essential" relevance of a tendency for the matrix. In a loose sense, the difference between potentiality and possibility can be understood in terms of cohesive and contextual matrixes, so long as one recognizes that these parameters are constantly shifting. To say that the distinction between possibility and potentiality is not based on temporal modes is of course not to say that they are in any sense "atemporal," but rather that in their nature as spatiotemporal dynamic tendencies they are alike spread throughout past, present, and future.

Perhaps one of the factors that lead to the attempt to distinguish potentiality and possibility by way of temporal modes is that we tend to use the term 'possibility' in relation to that which is consistently thinkable, as a result giving to possibility a sense of the "eternal" or the "ephemeral," while we do not utilize the term potentiality in the same way. We may say that if something is consistently thinkable it is possible, but we do not say that if something is consistently thinkable it is potential. The use of the term 'potentiality' seems linked much more closely with the unfolding of spatiotemporal concreteness, and hence its association with a concrete past. Yet dynamic possibility rather than conceptual possibility is distinguishable from potentiality only by its relevance to a particular matrix, and the "boundaries" between possibility and potentiality relative to a matrix are flexible and changing.

Moreover, the term 'possible' itself has a conceptual-dynamic interrelation that can further lead to confusion. In addition to the distinction between consistently thinkable abstract epistemic possibility and dynamic spatiotemporal possibility, there is needed a distinction within the consistently thinkable. In addition to logical conceivability or the absence of self-contradiction, there is what might be called the factually possible, which is the relation that dynamic possibility has to our body of information as contoured by the horizons of meaningfulness. What is factually possible must be logically possible. The reverse, however, does not hold; logical possibility does not imply factual possibility. But, if a particular assertion that is logically possible but not factually possible is combined with the body of available knowledge, a logically impossible set results.

Thus, at any time, a range of what is factually possible may be logically determined, though what specific possibility will in fact be actualized in the future can of course not be determined in this manner. Again, then, possibilities appear to take us away from the dynamic unfolding reality. Conversely, potentiality seems to carry with it connotations of a dynamic unfolding or power that eludes our epistemic sense of possibility. But to confer upon the "eternal" or "ephemeral" nature of abstract epistemic conceptual possibility a status constitutive of the ongoing unfolding of spatiotemporal reality is to confound an epistemic function with a metaphysical process.[39]

Moreover, even the "eternal" nature of abstract epistemic possibility is eternal only relative to a conceptual ordering, for what is possibile in our experience is in part a function of "what is there" and in part a function of the conceptual net we use to get hold of what is there, for reality as indeterminate is structurable in alternative ways, though it also constrains these ways, accommodating some as workable but not others. This statement brings us to the concept of world as the horizon for our understanding of processive concreteness.

Our perceptual world, which grounds the emergence of facts and objects and provides the foundation for other, more abstract or more imaginative worlds, is ontologically one with the spatiotemporal unfolding of an indeterminate reality. It is, metaphysically, that independently real. Yet a world is dependent upon the meaning system or interpretive context that grasps in a way in which the full concreteness of spatiotemporal reality is not, for a world is that perspective of the indefinitely rich reality that has been "fixed" or "carved out" by a system of meanings. Knowledge is abstractive and selective. A world, though concrete, is nonetheless selective in the sense that a world, as the concrete content denoted by a system of meanings, is a way in which the indeterminate concreteness of reality can be delineated or "fixed." A system, once chosen, limits the alternatives possible within it, but alternative systems may be possible. The indefinite richness of intermeshing matrixes is there. Where the conceptual "cuts" are imposed is, in part, our decision. A world is delineated by a system of facts, but facts are not independent of the selective knowledge process, for facts are abstracted portions of interacting continuities.

A world is by definition consistent because a world is the concrete content that is delineated by a interrelated network of consistent meanings. The world answers to the laws of excluded middle and noncontradiction, and thus it represents the ideal of that which can be conceptually articulated—and hence made precise—to its ideal limit. Perhaps it can be said, somewhat metaphorically, that while reality is the infinity of a continuum or ongoing process, the world is the logical fixation of an infinite number of possible cuts within it. The world is the context of meaning within which all other frameworks and objects may be articulated in the sense that the world is the "outermost" content or encompassing frame of reference or field of interest of organism-environment interaction, and hence of the conceptual cuts that can delineate experience consistently within the context of meaningful rapport. And, of course, as stressed in previous chapters, the "logical fixation" of possible cuts emerges within prereflective behaviorial attitudes as regulative and anticipatory.[40]

Such a world then, opens in one direction toward the structures of a processive, concreteness universe and the possibilities it presents, and in the other direction toward the structures of our modes of grasping it and the possibilities such modes of grasping allow. What can occur in the world must conform to the possibilities available within the world we have structured—though the world we have structured has arisen through successful interaction with the possibilities offered by the dynamic unfolding of spatiotemporal reality. Our interpretations are, as Derrida holds, rooted in the flux, but the flux is not "free-floating"; the flux is the flux of a temporally indeterminate universe in its ongoing dynamics, an indeterminate universe that renders senseless, indeed literally so, the concept of "mirroring" to which Rorty rightly objects or the supposed grasp of unmediated, decontextualized facts to which Derrida rightly objects.

Throughout many levels, truth is both made and found. The so-called tensions between truth as made and truth as found, between truth as changing and truth as fixed, result from focusing on diverse dimensions operative within the intertwining of human interpretive activity and the temporal unfolding of processive concreteness. We create the interpretive frameworks within which beliefs can emerge and be found true or false and within which investigation can tend toward agreement. The creative intelligence involved in radical changes and shifts of interpretive frameworks is influenced by sociocultural conditions, but is ultimately founded not in a relativistic, perspectivally closed historicism, but in an ontologically grounded, perspectivally open temporalism. The criterion for adequately cutting into the indefinitely rich matrix of possibilities of experience offered by the dynamics of an indeterminately rich concrete universe is workability, but workability can be established only relative to some meaningful network by which experience is "caught." Reality answers our questions and determines the workability of our meanings, but what answers it gives are partially dependent on what questions we ask, and what meaning structures work are partially dependent upon the structures we bring.

Intelligibility and truth, then, no more than the arrow of time, requires either the ontologically discrete or the ontologically determinate. As argued previously, the choice between temporal reversibility or a fixed past is a false set of alternatives ultimately rooted in temporal atomicity and the alternatives this brings with it, and is perhaps an unavoidable choice unless the attempt to find the fully determinate within spatiotemporal concreteness is abandoned. For the directional flow of possibilities and potentialities that provide for the arrow of time is tied to the indeterminate nature of past, present, and future. It is tied also to possibilities and potentialities as neither abstract nor fixed, but as concrete dynamic tendencies continually asserting their directional presence in the ongoing adjustments of temporal passage. This in turn grounds the possibility of the structure of interpretation, grounds the possibility of the structure of experience as experimental in the primal sense of temporal flow.

Neither the ceaselessly "becoming other" of reality nor its inherent indeterminacy leads to unintelligibility or the end of metaphysics. The postmodern tendency to so relate the two stems from the refusal to separate intelligibility from discreteness and fixity. The reality of the continuity of becoming other and the indeterminacy this brings with it provides for rational discourse and ongoing inquiry that is rooted in and provides perspectival knowledge about reality, so long as knowledge is not understood as a direct, uninterpreted seizure of what immediately "is," and truth is not understood as conformity or correspondence to the fixed discretes of a fully determinate reality. The choice between the groundlessness of Derrida's play or Rorty's conversation on the one hand and the grasp of reality in its "pristine purity" on the other is very different from the choice between the "groundlessness" of temporal reversibility and a fixed past, but these two different problematics are joined together by the common assumption that without a "place" for the fully determinate, the groundless alternative wins out.

Abstract knowledge claims do not constitute our main access to the natural universe; concrete experience does. The flux of life as it concretely occurs contains already a phenomenological dimension of human thrown-outness onto the universe through a primordial sense of temporal flow and the living meanings constitutive of the nature of experience as experimental with which such a primordial sense is interwoven. The being of humans in the natural universe and the knowing by humans of the natural universe are inseparably connected within the temporal structure at the heart of experience. And nowhere is this more manifest than in the temporal dynamics of the lived body-self.

Our access to the temporal pulsations of reality, then, is not to be found through reflective dismemberments of time, but rather these dynamic pulsations are manifest to us in our primal lived experience, experience in which our lived time consciousness, our sense of temporal directional flow, pulsates in harmony with the pulsations of the universe at large. These temporal pulsations of reality are part and parcel of our ongoing praxis that manifests throughout the temporal textures and directional flow of the universe at large.

This temporal universe in which we have our being is one in which our experience of temporality is at home and with which our temporally grounded activity is continuous; one in which our interpretive networks embodying dispositionally generated noetic possibilities can grasp, within the temporal flow, real dynamic tendencies of reality to produce operations of a certain type with a certain regularity; a universe, in brief, whose temporal features are embodied within the temporally pervasive textures that are at the heart of all experience and make possible even the very abstract conceptions that in turn work to deny the reality of these temporal features.

It has been seen that the characteristics of this temporal universe involve not perishing but change within the enduring; not succession but novelty emerging within dynamic continuity; not an atomicity of actualities but the emergence of quasidiscretes, be it present, past, and future, or persisting "things." Pragmatism thus attempts to account, in its own way, for a resisting, thick, ontologically dense, causally efficacious universe characterized by both persistence and change, continuity and discreteness, lawfulness and spontaneity, conformity and creativity, all of which enter into its creative advance.

This pragmatic position rules out all attempts to explain time in a way that "explains away" our sense of lived time as not ultimately a sense of cosmic time, either because there is no "real time" or because "real time" becomes self-contradictory; it rules out as well attempts to account for real time in a way that alienates it from our sense of lived time, either by turning to discrete units of atomic moments that cause but are unlike our sense of lived time, or by holding that lived time provides a seeming continuity composed ultimately of atomic durations.

But to accept the pragmatic offering of the temporal reintegration of humans and the cosmos at large via our primordial experience of temporal flow and the features contained therein requires relinquishing a long tradition of ingrained habits of thinking, habits of thinking that require a bedrock of the fully fixed, the fully final, the fully determinate, the atomic bit; that manifest the preference for the reified abstractions of reflective thought over the deliverences of our lived concrete attunement to and participation in the temporal textures of the universe at large in its ongoing directional flow. Temporality is so easily lost by reflective awareness because it is conveyed in the holistic nature of experience. This holistic nature of experience is part and parcel of the holistic nature of time. And the holistic nature of temporal movement, if it is to accommodate novelty in its continuous unfolding, brings with it an indeterminacy that is spread throughout. Only if we duly relinquish the last remnants of dismembered temporal parts resulting from a long enamorment with atomicity can temporal process have its due say.

NOTES

INTRODUCTION

1. See Richard Gale's article, "From the Specious to the Suspicious Present: The Jack Horner Phenomenology of William James," *Journal of Speculative Philosophy* 11 (1997).

2. Ilya Prigogine and Isabelle Stengers, *Order Out of Chaos* (New York: Bantam Books, 1984), pp. 214, 194. See also Brian Martine, "Mediation, Immediacy, and Time," *The Journal of Speculative Philosophy* 1.4 (1987).

3. This is in essence the intent utilized in my book, *Speculative Pragmatism* (Amherst: University of Massachusetts Press, 1986; paperback ed. Peru, Ill.: Open Court, 1990). In that work I have argued that the positions of the five major classical American pragmatists—Charles Peirce, William James, John Dewey, G. H. Mead, and C. I. Lewis—form a systematic and unified movement.

1. WHITEHEAD, WEISS, AND SPECULATIVE METAPHYSICS OF TIME

1. While it will be held that Neville is ultimately Whiteheadian in his understanding of time, yet his work is pervaded in very deep ways by the spirit and doctrines of American pragmatism.

2. *Concept of Nature* (London: Cambridge University Press, 1920), p. 12. Whitehead also rejected Bergson's claim that spatialization is a distorting form that science imposes on its objects. While this is one of the important issues between Whitehead and Bergson, F. S. C. Northrop has surely gone too far in characterizing it as the only really major point of disagreement between Whitehead and Bergson. "Whitehead's Philosophy of Science," *The Philosophy of Alfred North Whitehead*, Library of Living Philosophers, ed. Paul A. Schilpp (New York: Tudor, 1951), p. 169. In any case, the issue of space is not the topic of concern here.

3. *The Principle of Relativity* (Cambridge: Cambridge University Press, 1922), p. 32ff.

4. Ibid.

5. In *Process and Reality: An Essay in Cosmology* (New York: Harper & Row, 1960).

6. *The Aims of Education and Other Essays* (New York, Macmillan, 1929), p. 237.

7. His atomic occasions are of course not to be conflated with the atoms of science, for these latter are themselves composed of relationships among atomic occasions. But these Whiteheadian occasions are nonetheless modeled after the atomism of science.

8. *Process and Reality*, p. 363.

9. *Science and the Modern World* (New York: Macmillan, 1967), p. 126.

10. Ibid., p. 177.

11. Victor Lowe, *Understanding Whitehead* (Baltimore: The Johns Hopkins University Press, 1962), p. 229.

12. Whitehead refers to this as a "conceptual prehension," but this is the extended use of conceptual in Whitehead's philosophy.

13. *Process and Reality*, p. 123.

14. Ibid., p. 32.

15. Ibid., p. 283.

16. Ibid., p. 215.

17. Donald Sherburne, *A Key to Whitehead's Process and Reality* (Chicago and London: The University of Chicago Press, 1966), p. 38.

18. *An Interpretation of Whitehead's Metaphysics* (New Haven, Conn.: Yale University Press, 1959), pp. 80–81.

19. Ibid., p. 38 Sherburne goes into some considerations that were possibly thought by Whitehead to mitigate the difficulty, but questions their adequacy. This concerns Whitehead's point that the analysis into phases of concrescence is "only intellectual" and involves some arbitrariness in marking off individual prehensions. But as Sherburne notes, his entire philosophy puts such emphasis on the separate and sequential nature of prehensions that the appeal to this analysis as intellectual only seems at odds with his general position (p. 39).

20. Robert Neville, *Eternity and Time's Flow* (Albany: State University of New York Press, 1993), pp. 39–40.

21. Ibid.

22. See, for example, Jorge Nobo, *Whitehead's Metaphysics of Extension and Solidarity* (Albany: State University of New York Press, 1986), p. 248.

23. V. C. Chappell, "Whitehead's Theory of Becoming," *Journal of Philosophy* 58 (1961): 516–28; reprinted in *Alfred North Whitehead: Essays on His Philosophy*, ed. George L. Kline (Englewood Cliffs, N.J.: Prentice Hall, 1963), pp. 70–80.

24. *Science and the Modern World* (New York: Macmillan, 1967), p. 177.

25. Lewis S. Ford, "Afterword: A Sampling of Other Interpretations," *Explorations in Whitehead's Philosophy*, ed. Lewis S. Ford and George L. Kline (New York: Fordham University Press, 1983), pp. 314–15.

26. *Science and the Modern World*, p. 10.

27. Ibid., p. 126.

28. Ibid.

29. *Process and Reality*, pp. 238, 103f.

30. Ibid., p. 103.

31. Jorge Nobo, *Whitehead's Metaphysics of Extension and Solidarity* (Albany: State University of New York Press, 1986).

32. Ibid., p. 208.

33. *Process and Reality*, p. 53.

34. Ibid.

35. *Whitehead's Metaphysics of Extension and Solidarity*, p. 220 (italics added).

36. Ibid., p. 223.

37. Ibid.

38. Ibid.

39. Patrick Hurley also takes objection to the use of contiguity to solve certain problems. Yet this objection does not lead him to continuity but back to the affirmation of the Whiteheadian atomicity of temporal units. Patrick Hurley, "Time in the Earlier and Later Whitehead," *Physics and the Ultimate Significance of Time* (Albany: State University of New York Press, 1986), p. 94. His interest in Whitehead here lies well beyond the scope of the present concern.

40. Donald Sherburne, "Whitehead without God," *Process and Christian Thought*, ed. Delwin Brown, Ralph James Jr., and Gene Reeves (Indianapolis: Bobbs-Merrill, 1971), pp. 320–23. Reprinted, with revisions, from *The Christian Scholar* 50 (1967): 251–72.

41. *Explorations in Whitehead's Philosophy*, p. 326.

42. Edward Pols, *Whitehead's Metaphysics: A Critical Examination of Process and Reality* (Carbondale and Edwardsville: Southern Illinois University Press, 1967), pp. 101–95.

43. *Modes of Thought* (New York: Macmillan, 1933), p. 73.

44. *Philosophical Review* 46 (1937): 170–77.

45. *The Philosophy of the Present*, ed. Arthur Murphy (LaSalle, Ill.: Open Court, 1959), pp. 43, 49.

46. Ibid., pp. 20–21.

47. The following features can best be found encapsulated in *Eternity and Time's Flow*, p. 71ff.

48. Ibid.

49. Ibid., p. 180.

50. *Eternity and Time's Flow*, pp. 112–13.

51. *Recovery of the Measure* (Albany: State University of New York Press, 1989), p. 181.

52. *Eternity and Time's Flow*, p. 96.

53. Ibid., p. 105.

54. Ibid.

55. Ibid., p. 101.

56. The following is a gross oversimplification of Neville's position here, but one that I think does not distort it and is adequate for the point being made.

57. Ibid., p. 266.

58. *Eternity and Time's Flow*, p. 92.

59. Ibid.

60. This focus on temporality should not obscure the fact that Neville's creative use of pragmatic themes in general offers a sensitive and highly perceptive rendering of the position and houses insights and new directions important for anyone interested in the pragmatic perspective.

61. This discussion does not involve any claim that Neville's appeal to eternity indicates some weakness in his position on time. Within the framework of his theological interests, such an appeal does not reflect the weakness of a position that must call on eternity to solve a philosophic puzzle, but instead the strength of a position that develops from the start with an intent to both draw upon and point to the need for eternity—now redefined in accordance with his radically processive approach. Nor does the fact that the pragmatists do not make this move to eternity point in itself to the issue at hand. Rather, the point is that Neville's structuring of his enterprise reflects throughout the Whiteheadian focus on the ultimacy of atomicity.

62. "Charles Peirce's 'One Contribution to Philosophy' and His Most Serious Mistake," *Studies in the Philosophy of Charles Sanders Peirce*, second series, ed. Edward C. Moore and Richard S. Robin (Amherst: University of Massachusetts Press, 1964), pp. 455–74.

63. Ibid., p. 460.

64. Ibid., p. 469.

65. Ibid.

66. Ibid., p. 461.

67. Ibid., p. 473.

68. Ibid., p. 467.

69. Ibid., p. 469.

70. Ibid.

71. *Collected Papers of Charles Sanders Peirce*, vols. 1–6, ed. Charles Hartshorne and Paul Weiss (Cambridge: Belknap Press of Harvard University, 1931–35); vols. 7 and 8, ed. Arthur Burks (Cambridge: Harvard University Press, 1958), 7:653.

72. Dewey, *The Philosophy of Alfred North Whitehead*, Library of Living Philosophers, ed. Paul Schilpp (New York: Tudor, 1951), pp. 654–56.

73. Ibid., p. 656.

74. Ibid.

75. The five finalities are substance, being, possibility, existence, and unity.

76. *Beyond All Appearances* (Carbondale and Edwardsville: Southern Illinois University Press, 1974), pp. 228–29.

77. *First Considerations* (Carbondale and Edwardsville: Southern Illinois University Press, 1977), p. 107.

78. Weiss refers to pragmatism frequently in his writings, usually in a highly critical vein, but never views the position as in any way related to process.

79. The following discussion is not concerned with Whitehead's position per se but only with Weiss's interpretation of it.

80. Weiss sees the Humean and Whiteheadian views of causality "belonging together" as two sides of the same coin. *Cinematics* (Carbondale and Edwardsville: Southern Illinois University Press, 1975), p. 68.

81. *Beyond All Appearances*, p. 53.

82. Ibid.

83. *First Considerations*, p. 178.

84. *Modes of Being*, 2 vols. (Carbondale: Southern Illinois University Press, 1958), p. 223.

85. Ibid., p. 242.

86. Ibid., p. 232.

87. Ibid.

88. Weiss here introduces his notion of the Dunamis and characterizes it in this way. *Privacy* (Carbondale and Edwardsville: Southern Illinois University Press, 1983), p. 17. His entire discussion of the Dunamis has been dismissed by some critics as superfluous.

89. Ibid., p. 21.

90. *Sport: A Philosophic Inquiry* (Carbondale and Edwardsville: Southern Illinois University Press, 1969), p. 115.

91. Ibid.

92. Ibid.

93. *Reality* (Carbondale and Edwardsville: Southern Illinois University Press, 1938), p. 233.

94. *Modes of Being*, 1:229.

95. *Man's Freedom* (New Haven, Conn.: Yale University Press, 1950), p. 18.

96. *History: Written and Lived* (Carbondale and Edwardsville: Southern Illinois University Press, 1962), p. 153.

97. *Modes of Being*, 1:116.

98. For his own statement of this progression see *First Considerations*, p. 204.

99. See *Past, Present, and Future: A Philosophical Essay about Time* (Urbana and Chicago: University of Illinois Press, 1991).

100. Ibid., p. 127.

101. Ibid.

102. Ibid., p. 124ff.

103. *Process Metaphysics: An Introduction to Process Philosophy* (Albany: State University of New York Press, 1996), p. 43 This is the case to such an extent that Rescher finds the position to house an objectionable anthropomorphism. Ibid.

104. Neville's position, involving an attempted synthesis of pragmatic and Whiteheadian thought as discussed earlier, is an attempt to unify the lived sense of time and its objective reality, but as also discussed earlier, it ultimately falls into the Whiteheadian camp of the ultimacy of discrete units of time.

2. DERRIDA, MCTAGGART, AND THE DEMISE OF TIME

1. See especially John Caputo, "The Economy of Signs in Husserl and Derrida: From Uselessness to Full Employment," in *Deconstruction and Philosophy: The Texts of Jacques Derrida*, ed. John Sallis (Chicago: University of Chicago Press, 1987), pp. 103–4.

2. "No element can function as a sign without referring to another element which itself is not simply present. This interweaving results in each 'element'—phoneme or grapheme—being constituted on the basis of the trace within it of the other elements of the chain or system." Jacques Derrida, *Posi-*

tions, trans. Alan Bass (Chicago: University of Chicago Press, 1981), pp. 26–27.

3. Hilary Lawson, *Reflexivity: The Post-modern Predicament* (LaSalle, Ill.: Open Court, 1985), p. 100.

4. This denial of presence reveals the priority of writing over speech. For speech is already bound into the network of signifiers, taking it beyond the so-called purity of its immediately present meaning. Instead of the meaning of a sign as immediately present, it is disseminated throughout the system of signifiers. Hence, since the priority that has been accorded to speech over writing, or phonocentrism, is a result of the view of presence and the present, with the denial of these, there is a concomitant denial of the priority of speech over writing.

5. Rodolphe Gasche "Infrastructures and Systematicity," in *Deconstruction and Philosophy: The Texts of Jacques Derrida*, pp. 11–12.

6. John Caputo, "The Economy of Signs in Husserl and Derrida," p. 105.

7. Rodolphe Gasche, "Infrastructures and Systematicity," p. 12.

8. Thus, it generates a kind of meaning in the very positioning of such words as "green is or." As John Caputo states, "the expressions 'green is or' and 'abracadabra' are but isolations of this structure and hence a welcome *liberation* from the rule of intuitionism, a liberation which is in fact made possible by the reduction of the rules of a priori grammar. . . . This reduction . . . *liberates* the signifier from the oppressive regime of intuitionism and its unfair demand that every signifier leads to Being, presence, objectivity, even when such demands cannot be met. Intuitionism exacts a tax which no one can pay." "The Economy of Signs in Husserl and Derrida," p. 105.

9. Although Derrida exploits this latent tension within Husserl's account between the "instant as a point" (*Speech and Phenomena: And Other Essays of Husserl's Theory of Signs*, trans. David B. Allison and Newton Bercer [Evanston, Ill.: Northwestern University Press, 1973], p. 60) and the thickness and depth of the present, he is aware that, in spite of this tension, Husserl himself is explicitly convinced that no "now" can be isolated as a pure instant, a pure punctuality. Ibid., pp. 61–62.

10. *Speech and Phenomena*, p. 60.

11. Ibid., p. 61.

12. Ibid., p. 65.

13. Ibid., p. 64.

14. Ibid., p. 65.

15. Ibid., p. 68.

16. Ibid., p. 142.

17. Ibid., p. 85.

18. Ibid., p. 66.

19. Ibid., p. 85.

20. Ibid., p. 86.

21. Ibid., p. 142.

22. Ibid., p. 147.

23. Ibid.

24. Ibid., p. 66.

25. Ibid., p. 146.

26. Derrida, *Of Grammatology*, trans. Gayatri Chakravorty Spivak (Baltimore: John Hopkins University Press, 1976), p. 69.

27. Ibid.

28. Jacques Derrida, *Positions*, trans. Alan Bass (Chicago: University of Chicago Press, 1981), p. 28.

29. Ibid., p. 29.

30. *Speech and Phenomena*, p. 85.

31. Ibid., p. 93.

32. From J. M. E. McTaggart, *The Nature of Existence*, vol. 2 (Cambridge: Cambridge University Press, 1927), book 5, chapter 33; reprinted in *The Philosophy of Time: A Collection of Essays*, ed. Richard Gale (Atlantic Highlands, N.J.: Humanities Press, 1978), p. 97.

33. Ibid., p. 88.

34. Richard Gale, *The Language of Time* (London: Routledge and Kegan Paul, 1968), p. 214.

35. M. M. Schuster, "Is the Flow of Time Subjective?" *The Review of Metaphysics* 39 (1986): 695–714.

36. Ibid., p. 697. Italics in original.

37. "A Structural Model for Temporal Passage," *The Southern Journal of Philosophy* 32 (1994): 1–18.

38. Robert S. Brumbaugh, *Unreality and Time* (Albany: State University of New York Press, 1984), pp. 14–15.

39. Ibid.

40. Ibid., p. 12.

41. See, for example, A. Grunbaum, "The Status of Temporal Becoming," *The Philosophy of Time*, ed. R. M. Gale (Atlantic Highlands, N.J.: Humanities Press, 1968), pp. 322–54; "The Meaning of Time," *Basic Issues in the Philosophy of Time*, ed. E. Freeman and W. Sellars (LaSalle, Ill.: Open Court, 1971), pp. 195–228; D. H. Mellor, *Real Time* (Cambridge: Cambridge University Press, 1984). For an excellent analysis of Mellor's position from a pragmatic perspective, see Gary Calore's "Pragmatism, Becoming, and the B-Theory of Time," presented at the March 1994 meeting of the Metaphysical Society of America, Vanderbilt University.

42. Errol Harris, *The Reality of Time* (Albany: State University of New York Press, 1988), p. 37.

43. Ibid., pp. 36–37.

44. Ibid., p. 165.

45. Ibid., pp. 48–51.

46. While the focus is on Whitehead, various of these reasons in varying degrees apply to the other philosophers discussed in chapter 1.

47. David Wood, *The Deconstruction of Time* (Atlantic Highlands, N.J.: Humanities Press, 1989), p. 311.

48. *The Wake of Imagination: Toward a Postmodern Culture* (Minneapolis: University of Minnesota Press, 1988), p. 359. One answer Kearney offers, for example, is that "when the deconstructionist asks 'who is this I?' it may well—perhaps unbeknownst to itself—serve the ethical purpose of de-centering the epistemological subject," opening it to an awareness of responsibility to the

other-than-self (pp. 351–52, note 3). Yet, given the temporal atomism and resulting absolute alterity that undergirds the Derridian position, the attempt to reach the other-than-self may house implicit but deeply embedded problems. This issue will be discussed further in chapter 6.

49. Richard Bernstein, "Serious Play: The Ethical-Political Horizon of Jacques Derrida," *Journal of Speculative Philosophy* 1 (1987): 112.

50. Ibid., p. 98.

51. As was noted in the previous chapter, Neville does hold that this prereflective sense throws us out onto a temporal universe with which it is congenial, but he does this by transferring to eternity the problem of obtaining true continuity from a conjoining of ultimately discrete units.

3. HEIDEGGER AND TEMPORAL EXISTENCE

1. *Heidegger's Pragmatism* (Ithaca and London: Cornell University Press, 1988).

2. *Heidegger's Pragmatism*, p. 280 (italics added to highlight point of contention). See Richard Rorty, *Essays on Heidegger and Others* (Cambridge: Cambridge University Press, 1991), pp. 38–39, n. 22. In "The Metaphilosophical Consequences of Pragmatism," *The Institution of Philosophy*, ed. Avner Cohen and Marcello Dascal (Totowa, N.J.: Rowman and Littlefield, 1988), Okrent offers an in-house criticism of some of Rorty's views and a defense of his own.

3. Daniel O. Dahlstrom, "Heidegger's Concept of Temporality: Reflections on a Recent Criticism," *The Review of Metaphysics* 44 (1995): 95–115, see pp. 96–97. Dahlstrom's explicit focus here is on the pragmatism of the Okrent-Rorty variety.

4. The complexities of Aristotle's view of time lie well beyond the scope of the present endeavor. What is of concern here is Heidegger's reading of Aristotle. For an interesting analysis of the Heideggerian and Derridian readings of Aristotle's writings on time and the resulting critique of Derrida that this involves, see J. Claude Evans, "Rigor and in Deconstruction," *Philosophy Today* 40 (1996): 55–60.

5. *What Is Called Thinking?* trans. J. Glenn Gray (New York: Harper & Row, 1968), p. 102.

6. It is called world-time because it has the character of significance. Heidegger, *Being and Time*, trans. John Macquarrie and Edward Robinson (New York: Harper & Row, 1962), p. 270.

7. Heidegger, *The Metaphysical Foundations of Logic*, trans. Michael Heim (Bloomington: Indiana University Press, 1984), p. 198.

8. *Being and Time*, 376.

9. Ibid., p. 170.

10. Ibid., p. 416.

11. *Being and Time* still allows for the existence of substances or essences other than the nonsubstantive human way of being, substances that ground the more abstract functional concepts of science (see p. 122). However, in the later

Heidegger the unity not just of second-level abstract objects of reflection but of the things of everyday experience lies in their functionality. See his analysis of the jug in "The Thing," *Poetry, Language, Thought*, trans. A. Hofstadter (New York: Harper & Row, 1971), p. 166ff.

12. See, for example, Heidegger, *Being and Time*, p. 122.

13. Heidegger, "Letter on Humanism," *Basic Writings*, ed. David Farrell Krell (New York: Harper & Row, 1977), p. 206.

14. Heidegger designates temporality (*Zeitlichkeit*) as the condition of the possibility of the constitution of the being of the human. However, the understanding of being belongs to this constitution of the human. Therefore, temporality must also be the condition of possibility of the understanding of being that belongs to the human. This Heidegger calls Temporality (*Temporalität*) with a capital T. Heidegger, *The Basic Problems of Phenomenology*, trans. Albert Hofstadter (Bloomington: Indiana University Press, 1982), p. 274ff.

15. Heidegger, *Being and Time*, p. 255.

16. "Letter on Humanism," *Basic Writings*, p. 216. (Reference to *Being and Time* on p. 255 of translated text.)

17. In a different though related fashion, it is precisely this "independent" aspect of the world as already there that Merleau-Ponty comes to develop explicitly in his later, more ontological considerations. Merleau-Ponty's later writings attempt to locate the phenomenological descriptions of his early work within an ontology that in no way negates, but rather broadens and rounds out the earlier phenomenology. In this process, the system of world, self, other selves, the role of the body, the view of truth, and the role of sensibility take on expanded ontological dimensions. As Sallis put it, this provides "an ontological rehabilitation of the sensible." John Sallis, *Phenomenology and the Return to Beginnings* (Pittsburgh: Duquesne University Press, 1973), p. 55; Merleau-Ponty, *Signs*, trans. Richard C. McCleary (Evanston, Ill.: Northwestern University Press, 1964), pp. 166–67.

18. The concept of extantness, of that which is independent of humans and their world, is not to be found in *Being and Time*. However, it is brought into clear focus in the early Heidegger in *The Basic Problems of Phenomenology* (see, for example, pp. 169, 286, 294–97). It is also found in *The Metaphysical Foundations of Logic*, trans. Michael Heim (Bloomington: Indiana University Press, 1984); See, for example, pp. 156–57.

19. *The Metaphysical Foundations of Logic*, p. 125.

20. A good example of this debate is to be found in "Heidegger's Concept of Temporality: Reflections on a Recent Criticism," in which Daniel Dahlstrom presents a detailed defense against Margot Fleischer's separation of authentic and original temporality and hence of her resulting criticisms of Heidegger. (Margo Fleischer, *Die Zeitanalysen in Heideggers 'Sein und Zeit': Aporien, Probleme und ein Ausblick* (Würzburg: Königshausen und Neumann, 1991).

21. See William Blattner's support of this claim, "Existential Temporality in *Being and Time*," *Heidegger: A Critical Reader*, ed. H. Dreyfus and H. Hall (Cambridge, Mass.: Blackwell, 1992), p. 99.

22. Graeme Nicholson, "Ekstatic Temporality in *Zein und Zeit*," *A Companion to Martin Heidegger's Being and Time*, ed. Joseph Kockelmans (Wash-

ington, D.C.: Center for Advanced Research in Phenomenology and University Press of America, 1986), p. 216.

23. Ibid., p. 244.

24. Yoko Arisaka, "Spatiality, Temporality, and the Problem of Foundation in *Being and Time*," *Phenomenology and Beyond: Selected Studies in Phenomenology and Existential Philosophy*, ed. John Caputo and Lenore Langsdorf, vol. 21, *Philosophy Today* 40 (1996): 40.

25. Heidegger, *Discourse on Thinking*, trans. John Anderson and E. Hans Freund (New York: Harper & Row, 1966), p. 66ff.

26. See, for example, his analysis of the jug in "The Thing," p. 166ff.

27. What Heidegger here means by *region* is not any one region, or regions in general, but "the region of all regions," that which gives rise to particular regions, including the human region. See Heidegger's *Discourse on Thinking*, trans. John M. Anderson and E. Hans Freund (New York: Harper & Row, 1966), p. 66, n. 1.

28. In Yoko Arisaka's "Spatiality, Temporality, and the Problem of Foundation in *Being and Time*," there is a brief focus on that-which-regions. But Ariskaska interprets this as being Heidegger's way of dealing with the issue of spatiality, and gives it no temporal significance.

29. Ibid., p. 58.

30. Heidegger, *On Time and Being*, trans. Joan Stambaugh (New York: Harper Torchbooks, 1972), p. 15.

31. Ibid., p. 85.

32. Ibid., p. 64.

33. Ibid., p. 67.

34. Ibid.

35. Ibid.

36. Heidegger, *On Time and Being*, pp. 19–22.

37. Ibid., p. 16.

38. Heidegger, *What Is Called Thinking?* See especially p. 113ff. "What 'calls for' thinking?" is the fourth and decisive way of asking the question, "What is called thinking?"

39. "Letter on Humanism," *Basic Writings*, ed. David Krell (New York: Harper & Row, 1977), p. 231.

40. Heidegger, *Being and Time*, p. 100.

41. Heidegger, *The Basic Problems of Phenomenology*, pp. 168–69.

42. Here it should be stressed that in speaking of scientific objects as a reflective abstraction from the everyday world of lived experience, or as a narrow view of the theoretical gaze, neither pragmatism nor Heideggerian thought intends this to indicate that the objects of science are merely abstracted from everyday experience by leaving aspects of it behind. There is a radical creativity involved in developing any scientific world that frequently leads its contents to contradict the sense of lived experience. Yet, however creative and strange to common sense the contents of science may become, the creativity emerges from the world of everyday experience and the theoretical content must be verified by a return to the things used in everyday experience, no matter how tenuous this link may become. This point will be further developed in the next chapter.

43. Martin Heidegger, *History of the Concept of Time*, trans. Theodore Kisiel (Bloomington: Indiana University Press, 1985), pp. 217–18.

44. *The Basic Problems of Phenomenology*, p. 169.

45. Ibid.

46. Ibid.

47. Ibid., p. 170.

48. Heidegger, *The Metaphysical Foundations of Logic*, trans. Michael Heim (Bloomington: Indiana University Press, 1984), p. 195.

49. Heidegger, "The Origin of a Work of Art," *Martin Heidegger, Basic Writings*, ed. David Farrell Krell (New York: Harper & Row, 1977). Elsewhere Heidegger refers to earth as part of the fourfold. There earth, as "under the sky" (which is also part of the fourfold) is taken in a more limited sense. Heidegger, "Building Dwelling Thinking," *Basic Writings*.

50. Ibid., p. 169.

51. Ibid., pp. 169–73.

52. Ibid., p. 175.

53. Ibid., p. 173.

54. Ibid., p. 178.

55. "Building Dwelling Thinking," p. 325.

56. "The Origin of the Work of Art," *Basic Writings*, p. 169.

57. "Building Dwelling Thinking," p. 326.

58. "The End of Philosophy and the Task of Thinking," *Basic Writings*, p. 391.

59. Ibid., p. 392.

60. Heidegger, *What Is Philosophy*, trans. William Kluback and Jean Wilde (New Haven, Conn.: College and University Press, 1958), p. 77.

61. Heidegger, *The Basic Problems of Phenomenology*, p. 327

62. *Radical Hermeneutics* (Bloomington: Indiana University Press, 1987), p. 177.

63. See Mead on this characterization, *The Philosophy of the Act*, ed. Charles Morris (Chicago: The University of Chicago Press, 1938), pp. 277–78.

64. John D. Caputo, *Demythologizing Heidegger* (Bloomington and Indianapolis: Indiana University Press, 1993), p. 125.

65. Ibid., p. 126.

66. David Wood, *The Deconstruction of Time* (Atlantic Highlands, N.J.: Humanities Press International, 1989).

67. See the discussions of these trends earlier in this chapter.

4. HEIDEGGER, SCIENTIFIC METHOD, AND TEMPORALITY

1. "Letter on Humanism," p. 231. He there links the "biologism" of pragmatism with its "scientism." See also the reference to this at the beginning of the previous chapter.

2. The term 'methodology' in science means something more limited for Heidegger than the present use of the term 'method'. Methodology, as used by

Heidegger, will be seen below to indicate one interrelated set of features of scientific method in general.

3. Causal connections are always expressed as relations among particular types of objects or events, and the nature of the events or objects being connected enters into the very understanding of the nature of the causal relationship sustained. This focus on scientific method as the method of causal analysis is thus still not purified of content and represents a lingering influence of modern worldview thought. This brief sketch of the distinction between scientific method and scientific content within pragmatic philosophy, as well as the following discussion of its understanding of scientific method, is examined and supported in some depth in *Speculative Pragmatism*.

4. G. H. Mead, "The Definition of the Psychical," *Mead: Selected Writings*, ed. A. J. Reck (New York: Bobbs-Merrill, 1964).

5. John Dewey, *Experience and Nature*, vol. 1 (1981), *The Philosophy of John Dewey: The Later Works*, ed. Jo Ann Boydston (Carbondale and Edwardsville: Southern Illinois University Press, 1981–), p. 37.

6. It is too easy to get entangled with problems of language at this point. Pragmatists tend to refer to the objects of science as a different kind of object than the objects of common sense. Heidegger, however, distinguished between objects and things, reserving 'things' for that which comes to presence within primary experience and 'objects' for the content of the abstract focus of science. Even here, however, there is a possible point of confusion. When Heidegger discusses the thing in his most famous work in this area (*What Is a Thing?*, trans. W. Barton and V. Deutsch [Chicago: Regnery Press, 1967]), he is dealing with the Kantian concept of thing, which corresponds most closely to his own use of 'object'. This Kantian concept of 'thing' is not appropriated by Heidegger in his own use of 'thing'. (See, for example, "The Thing", "The Turning", in *The Question Concerning Technology and Other Essays*, trans. William Lovitt [New York: Harper & Row, 1977].) The following essay will speak of the objects of science and the things of primary or foundational experience. It makes use of Heidegger's own distinction in use of language in a way that is not distortive of the pragmatic position, which clearly distinguishes the content of science from the content of the everyday world of perception, and which stays within the bounds of possible commonsense renderings of these terms, thus avoiding an overly technical language.

7. William James, *The Principles of Psychology*, 2 vols. (1981), *The Works of William James*, ed. Frederick Burkhardt (Cambridge: Harvard University Press, 1975–), 2:1232–34.

8. Dewey, *The Quest for Certainty*, vol. 4 (1984), *The Later Works*, pp. 163–65.

9. Heidegger, "The Age of the World Picture," *The Question Concerning Technology and Other Essays*, trans. William Lovitt (New York: Harper & Row, 1977), p. 118.

10. It is sometimes held that the concept of the scientific object as "decontextualized," found in the early Heidegger, conflicts with the scientific world characterized by the later Heidegger. However, though the objects of science are decontextualized from the everyday world, they are done so in conformity with

the possibilities of a projected scientific world. Hence, Heidegger refers to the world of physics as the "unworlded world" (Heidegger, *History of the Concept of Time*, pp. 217–18), for it has lost the concrete dimensions of the world in which we dwell.

11. Heidegger, "The Age of the World Picture," pp. 117–20.

12. For example, see James, *Principles of Psychology*, 2:961; Peirce, *Collected Papers*, vols. 1–6, ed. Charles Hartshorne and Paul Weiss (Cambridge: Belknap Press of Harvard University, 1931–35); vols. 7 and 8, ed. Arthur Burks (Cambridge: Harvard University Press, 1958), 7:498 (hereafter cited using only two-part conventional notation).

13. Heidegger, "The Age of the World Picture," pp. 120–24. It was indicated earlier that Heidegger's use of 'methodology' is more limited than 'method' as used in this chapter. Scientific method, as used in this essay, is best expressed by Heidegger's use of the term 'research', which incorporates projection and its related rigor, methodology, and ongoing activity, with each mutually requiring the other.

14. "The Age of the World Picture," p. 120.

15. Ibid.

16. Ibid., p. 121.

17. Peirce, 5:384.

18. "The Age of the World Picture," p. 124.

19. Ibid., appendix 2, p. 138.

20. Dewey, "Does Reality Possess Practical Character?" vol. 4 (1977), *The Philosophy of John Dewey: The Middle Works*, ed. Jo Ann Boydston (Carbondale and Edwardsville: Southern Illinois University Press, 1976–83), pp. 137–38.

21. Dewey, *Essays in Experimental Logic* (New York: Dover, 1916), p. 86.

22. Peirce, 5:181. Peirce's technical term for such creativity activity is abduction. This shading of scientific abductions into everyday perceptual claims is not a continuity of content organized but of method of organization.

23. G. H. Mead, *Philosophy of the Act* (Chicago: University of Chicago Press, 1938), p. 25.

24. C. I. Lewis, *Mind and the World Order* (New York: Dover, 1929), appendix A, esp. pp. 395–97.

25. For a good analysis of this point, see Dewey's discussion of "the table," *The Quest for Certainty*, pp. 189–90.

26. Science is detached from our world of commonsense engagement because its *objects* are detached abstractions from it. Scientists, however, are significantly and actively engaged with their world of scientific objectivities, though such engagement is more deliberately controlled and more narrowly focused.

27. This richer matrix is represented in James's understanding of pure experience and Dewey's similar understanding of a primordial experience as a processive continuity of activity undifferentiated by any thought distinctions. Here it should perhaps be stressed that such a level does not close one within the subjective but represents the matrix of a primordial interactive unity within the richness of an ontological embeddedness.

28. Heidegger, "The Age of the World Picture," pp. 218–19.

29. Ibid., pp. 119–20.

30. *Being and Time*, p. 195.

31. See, for example, Heidegger, *Being and Time*, p. 122.

32. Thus Heidegger can emphasize a certain essential "powerlessness" of the human mode of being (*The Metaphysical Foundations of Logic*, p. 215).

33. Heidegger, "The Age of the World Picture," p. 122. Though Heidegger, in viewing understanding as a structure of the human way of being, is concerned with focusing on understanding in a nonspectator fashion, he does not tend to make explicit reference to the spectator/nonspectator debate. Such an explicit reference can be found, however, in *The Basic Problems of Phenomenology*, p. 276.

34. See p. 122.

35. See his analysis of the jug in "The Thing," p. 166ff. A parallel understanding of this functionality within pragmatic philosophy was given earlier.

36. "The Age of the World Picture," p. 120.

37. "The Turning," p. 42. In this passage Heidegger is explicitly referring to the danger of the "enframing" of technology, but it applies equally well, perhaps even more appropriately, in the present context.

5. THE TEMPORAL A PRIORI

1. I have explored the implicit development of an a priori in Peirce's philosophy in *Charles Peirce's Pragmatic Pluralism* (Albany: State University of New York Press, 1994), chapter 2, and have focused on the importance of Lewis's concept of the a priori for pragmatism in general in my *Speculative Pragmatism*, chapter 3.

2. Heidegger, *Being and Time*, see especially pp. 87, 250. Further, Heidegger, in his concern to avoid the primacy of consciousness or subjectivity, is reacting against Husserl's phenomenology with its priority of knowing.

3. See *Mind and the World Order* (New York: Dover, 1956), appendix D, and *An Analysis of Knowledge and Valuation* (LaSalle, Ill.: Open Court, 1962), introduction.

4. *The Basic Problems of Phenomenology*, p. 276.

5. As Lewis states, a meaning is "a sort of purposive attitude," *Mind and the World Order*, p. 228.

6. "Letter on Humanism," *Basic Writings*, p. 231.

7. As Heidegger so well clarifies the meaning of constituting, it "does not mean producing in the sense of making and fabricating; it means letting the entity be seen in its objectivity." *History of the Concept of Time*, p. 71.

8. This point is developed by me in some detail in relation to Lewis and to classical American pragmatism in general in "Pragmatic Meaning and the Phenomenological Perspective: Some Common Denominators," *The Journal of Speculative Philosophy* 1 (1987), and to Lewis exclusively in "Lewis' Pragmatic Kantianism: Toward Dialogue with Phenomenology," *Phenomenology: Dialogues and Bridges*, ed. Ronald Bruzina and Bruce Wilshire (Albany: State University of New York Press, 1982).

9. C. I. Lewis, "Autobiography," *Collected Papers of Clarence Irving*

Lewis, ed. John D. Goheen and John L. Mothershead Jr. (Stanford, Calif.: Stanford University Press, 1970), p. 18.

10. Lewis, "Replies to My Critics," *The Philosophy of C. I. Lewis*, ed. P. A. Schilpp, Library of Living Philosophers series (LaSalle, Ill.: Open Court, 1968), p. 660.

11. For Heidegger understanding, rather than an act, is a mode of being of Dasein.

12. *Basic Problems of Phenomenology*, p. 70.

13. Lewis, *An Analysis of Knowledge and Valuation*, p. 361.

14. C. I. Lewis, *An Analysis of Knowledge and Valuation* (La Salle, Illi.: Open Court, 1946), p. 110.

15. Lewis, *An Analysis of Knowledge and Valuation*, p. 17.

16. *Dewey, Logic: The Theory of Inquiry, The Later Works*, ed. Jo Ann Boydston (Carbondale and Edwardsville: Southern Illinois University Press, 1986), 12:17.

17. Ibid.

18. By the term 'specious present' Lewis does not mean a false present but rather a spread-out present.

19. Lewis, *Mind and the World Order* (New York: Dover, 1956), p. 58.

20. See, for example, Heidegger, *Being and Time*, p. 122.

21. The nature of schemata in the philosophies of Heidegger and Lewis is here just briefly sketched, but it is examined in some detail, though in a much more limited context, in an article co-authored by myself and Patrick Bourgeois, "Schemata: Heidegger and Lewis," *International Philosophical Quarterly* 27 (1987).

22. Heidegger objects that Kant failed to link the process of temporal formation to its foundation in primordial time.

23. Clive Cazeaux has observed, perhaps correctly, that Heidegger's coinage of the term "schema-image" to capture the essentially image-producing nature of a schema may obscure as much as it clarifies. "Metaphor and Heidegger's Kant," *The Review of Metaphysics* 49 (1995): 359.

24. Heidegger, *Kant and the Problem of Metaphysics*, p. 102.

25. Ibid., p. 101.

26. Ibid., p. 100.

27. Ibid., p. 101.

28. This phenomenalistic interpretation can be seen throughout *The Philosophy of C. I. Lewis*, Library of Living Philosophers, ed. Paul Schilpp (LaSalle, Ill.: Open Court, 1968).

29. Lewis, *An Analysis of Knowledge and Valuation*, p. 134.

30. Ibid., p. 134.

31. Although Lewis usually speaks of sense meaning as a precise, explicit schema, yet sense meaning is, for Lewis, intensional or conceptual meaning, and this he frequently identifies as a disposition or habit. He clarifies this dual aspect of sense meaning when he observes that "a sense meaning *when precise and explicit* is a schema." *An Analysis of Knowledge and Valuation*, p. 134.

32. By 'intuition' Heidegger means only simple apprehension of what is found, and corresponds to "seeing" in the broad sense of the term. *History of the Concept of Time*, p. 47.

33. Ibid., p. 72

34. See Lewis, *Mind and the World Order*, pp. 25–26, for his own statement of this unity of reciprocity. This is implicit throughout his entire criticism of rationalism and traditional empiricism that runs throughout the book.

35. *History of the Concept of Time*, pp. 74–75.

36. *Basic Problems of Phenomenology*, p. 324.

37. *Mind and the World Order*, pp. 230–31.

38. See *History of the Concept of Time*, pp. 324–25.

39. *Mind and the World Order*. See especially chap. 9.

40. "A Pragmatic Conception of the A Priori," *Collected Papers of Clarence Irving Lewis*. Alternative a priori delimitations of course provide the basis for Heidegger's approach to the "historicity of being."

41. Though the meanings change, words used to designate them may remain the same.

42. Peirce, 8:191, 2:29.

43. For Lewis, "Meaning cannot be literally put into words or exhibited by exhibiting words and the relations of words." *An Analysis of Knowledge and Valuation*, p. 140.

44. Ibid., p. 133.

45. *Basic Problems of Phenomenology*, pp. 195–97. Heidegger points out that though Mill's classification does not agree with Kant's, this difference is irrelevant here. Ibid., p. 195.

46. Lewis, *An Analysis of Knowledge and Valuation*, pp. 161–63.

47. Ibid., p. 163.

48. For a detailed support of this claim, see my *Charles Peirce's Pragmatic Pluralism*, chapter 2.

49. *Collected Papers*, 4:233.

50. *Logic: The Theory of Inquiry*, ibid., p. 243.

51. The way this entire process operates in the area of abstract formal systems is discussed in some depth in my article, "Logic and Ontological Commitment: Lewis and Heidegger," *The Journal of Speculative Philosophy* 9 (1995).

52. James, *The Principles of Psychology*, 2:1255, footnote.

53. Ibid.

54. The ambiguities of James's understanding of the a priori as found in the *Principles* falls into focus within the context of the above discussion. James holds that there is a coerciveness among necessary and eternal relations that the mind finds between certain of its ideal conceptions, and that they form a determinate system independent of the order of frequency in which experience may have associated the conception's originals in time and space. James, *The Principles of Psychology*, 2:1255. So far as some of nature's realities fit this network, "we can make a priori propositions concerning natural fact" (ibid., p. 1269). Yet, he also indicates that our regulative principles are themselves conditioned by the experiences which they serve to organize (ibid., pp. 1215–70). Again, he holds at times that there is a sharp distinction between the way in which we justify the a priori truths of logic and mathematics on the one hand and the a posteriori truths of physics on the other (ibid.). Yet there are places where James comes close to holding that physical theories and theories of

mathematics are alike not only in being "spontaneous variations" but also in being "rational propositions" (ibid.). What seems implicit here is not an ambiguous vacillation but a development toward a view that physical truths contain an a priori element, though for physical truths, such an a priori element both arises within the matrix of experience and must be found workable within the context of experience.

6. THE TEMPORAL SELF

1. *The High Road around Modernism* (Albany: State University of New York Press, 1992), pp. 99–100.
2. "The Social Self," *Mead: Selected Writings*, p. 144ff.
3. Ibid., p. 142.
4. Ibid.
5. *Mind, Self, and Society*, ed. Charles Morris (Chicago: University of Chicago Press, 1934), p. 174.
6. "The Social Self," p. 148.
7. Ibid.
8. "The Mechanism of Social Consciousness," *Mead: Selected Writings*, p. 141.
9. Ibid.
10. Ibid.
11. *Mind, Self, and Society*, p. 173.
12. Ibid., p. 177.
13. *Mind, Self, and Society*, p. 136.
14. *The Individual and the Social Self: Unpublished Work of George Herbert Mead*, ed. David Miller (Chicago: University of Chicago Press, 1982), p. 148.
15. *Mind, Self, and Society*, 372.
16. Ibid., p. 353.
17. Ibid., p. 371.
18. Ibid., p. 178.
19. Ibid., pp. 174–75. Mitchell Aboulafia assimilates the self to the "me" pole, holding that while Mead often refers to the "I" and "me" as self, in other contexts the self is basically the "me." *The Mediating Self: Mead, Sartre, and Self-Determination* (New Haven, Conn.: Yale University Press, 1986) (see p. 18, n. 18, of his book for his specific statement of this intended use.) As here indicated, however, the self is in a certain sense most identified with the lived intentional body or the biologic dimension in its aspect of creativity, a function identified with the subject "I" rather than the object "me."
20. *Mind, Self, and Society*, p. 174, n.
21. Ibid., p. 176.
22. Ibid.
23. *The Philosophy of the Act*, ed. Charles Morris (Chicago: University of Chicago Press, 1938), pp. 65–66.
24. *Mind, Self, and Society*, pp. 10, 177.

25. Ibid., p. 178.
26. Ibid., p. 174.
27. This concept will be discussed more fully in the following chapters.
28. *Meaning and Action: A Critical History of Pragmatism* (New York: Bobbs-Merrill, 1968), p. 259.
29. Norbert Wiley, in "Peirce, Mead and the Internal Conversation," a paper presented at the American Sociological Association, August 1989, makes a distinction between the "me" as past and the "me" as object, but in so doing divorces the "me" as object from temporality.
30. "The Social Self," p. 142.
31. *Mind, Self, and Society*, p. 151.
32. Ibid., p. 376. Emphasis added.
33. *Mind, Self, and Society*, pp. 177–78. Emphasis added.
34. Ibid., p. 140.
35. Ibid., p. 176.
36. While Mead at times explicitly distinguishes between social as intrasystematic and sociality as intersystematic, he does not rigorously adhere to this distinction in terminology.
37. *The Philosophy of the Present*, ed. Arthur Murphy (LaSalle, Ill.: Open Court, 1959), p. 77. This distinction lies implicit in Hans Joas's assertion that "with the concept of 'sociality' Mead emphatically does not mean a general systematic character, but rather the stage of the origination of something new in a present." *G. H. Mead: A Contemporary Reexamination of his Thought* (Cambridge: MIT Press, 1985), p. 183.
38. Abdoulafia finds an ambiguity in Mead's use of the "I," claiming that Mead views it on the one hand as merely a functional distinction, while on the other hand he views it as the source of real novelty and as rescuing us from determinism (*The Mediating Self: Mead, Sartre, and Self-Determination* [New Haven, Conn.: Yale University Press, 1986], p. 25). Aboulafia seems to identify functional and fictional. Aboulafia's generally keenly perceptive analysis runs into this kind of problem with Mead because his way of focusing on Mead in relation to Sartre draws him away from any concern with the centrality of the lived body.
39. Ernst Tugendhat, *Self-Consciousness and Self-Determination*, trans. Paul Stern (Cambridge: MIT Press, 1986), pp. 234–48.
40. Jürgen Habermas, *The Theory of Communicative Action*, vol. 2, trans. Thomas McCarthy (Boston: Beacon Press, 1987), p. 106.
41. "The Social Self," p. 145.
42. See, for example, Abdoulafia, *The Mediating Self: Mead, Sartre, and Self-Determination*, p. 24.
43. *Mind, Self, and Society*, p. 136.
44. See Mead, *The Philosophy of the Present*, p. 30.
45. *Mind, Self, and Society*, p. 135.
46. Ibid., p. 177. Because this minimal temporal I-me adjustment does not involve the level of active self-engagement that makes the me an object of reflection, Mead has no way to constructively develop this claim in his understanding of the self.

47. As has been previously stressed, even reflective awareness as actively engaged in the dialogical stretch of the temporal present is nonthematic. This is very different from the second-level thematization of the "me" in "self-observation."

48. Mead seems to recognize this at times. See, for example, *The Individual and the Social Self*, p. 148, and *Mind, Self, and Society*, p. 163. But the significance of this never seems to be fully recognized by him.

7. TEMPORALITY AND COMMUNITY

1. "Is Patriotism a Virtue?" Lindley Lecture, University of Kansas, 1984, pp. 19–20.

2. *A Theory of Justice* (Cambridge, Mass.: Harvard University Press, 1971).

3. "Justice as Fairness," *Philosophy and Public Affairs* 14 (1985): 239, n.

4. Ibid., p. 238.

5. *John Rawls, Political Liberalism* (New York: Columbia University Press, 1993).

6. Ibid., see, for example, pp. 22–28; 208; 242, n.

7. The legislators of justice can determine whether the implementation of any given set of principles at any given time is feasible, but Rawls thinks it hard to imagine that we do not now have all knowledge needed for the feasibility test. See *A Theory of Justice*, p. 137.

8. Rawls, *Justice as Fairness*, pp. 223–51.

9. Ibid., p. 15. In *Political Liberalism*, pp. 147–48, Rawls holds that an overlapping consensus is quite different than what he calls a "modus vivendi" as a compromise position between individuals or states at odds in aims or interests. However, this distinction is not relevant to the points under discussion.

10. *After Virtue*, 2nd ed. (Notre Dame, Ind.: University of Notre Dame Press, 1984), see especially chapter 15.

11. Ibid., p. 191.

12. Ibid., p. 222.

13. Ibid., p. 221.

14. Richard Rorty, *Consequences of Pragmatism* (Minneapolis: University of Minnesota Press, 1982), p. 173. Rorty here creates a dialogue between the pragmatist—which he considers himself to be—and the defender of the Enlightenment or the traditional philosopher.

15. Ibid., pp. 173–74.

16. *Contingency, Irony, and Solidarity* (Cambridge: Cambridge University Press, 1989), p. 41.

17. Ibid., p. 43.

18. Ibid., p. xvi.

19. These characterizations arise in his discussion of metaphor in relation to the contingency of language (ibid., p. 18), but in his discussion of the contingency of the self in the following chapter, it can be seen that what holds of metaphor holds as well for imagination. Ibid., p. 36.

20. Ibid., p. 17. This capriciousness can be seen in Rorty's assertion that "the difference between genius and fantasy" is that the former case represents "idiosyncrasies which just happen to catch on with other people." Ibid., p. 37.

21. The term 'pole' is intended to stress that the individual and the generalized other exist only within the context of community. Without the individual and the generalized other, there is no community, but without the ongoing interaction of adjustment or accommodation that is the very essence of community, there is no individual or generalized other.

22. A person may be a member, of more than one community, for there are diverse levels and types of communities. Any community consists of many subgroups, and though individuals may feel alienated from a particular community, they cannot really be alienated from society in general, for this very alienation will only throw them into some other community. *Mind, Self, and Society*, pp. 353–54.

23. Dewey, *Ethics*, *The Middle Works*, ed. Jo Ann Boydston (Carbondale and Edwardsville: University of Southern Illinois Press, 1976–83), vol. 5 (1978), p. 327.

8. TEMPORALITY AND INTERPRETIVE ACTIVITY

1. David Weissman, *Hypothesis and the Spiral of Reflection* (Albany: State University of New York Press, 1989), p. 51.

2. Ibid.

3. Paul Weiss, *Reality*, (Carbondale and Edwardsville: Southern Illinois University Press, 1938), pp. 45–46.

4. Weiss, *Beyond All Appearances*, (Carbondale and Edwardsville: Southern Illinois University Press, 1974), p. 44.

5. Robert Neville, *Recovery of the Measure*, (Albany: State University of New York Press, 1989), pp. 50–51.

6. Bertrand Helm's book, *Time and Reality In American Philosophy* (Amherst: University of Massachusetts Press, 1985), is an exception, and his book in relation to this work perhaps helps reveal how important an issue this is to interpreting the pragmatists. For our differing understandings concerning the function of time in various philosophers leads to radically diverse interpretations of some of their key doctrines.

7. *Collected Papers of Charles Sanders Peirce*, vols. 1–6, ed. Charles Hartshorne and Paul Weiss (Cambridge, Mass.: Belknap Press of Harvard University, 1931–35); vols. 7 and 8, ed. Arthur Burks (Cambridge, Mass.: Harvard University Press, 1958), 6:86.

8. 6:103.

9. Carl Hausman's excellent and highly perceptive analysis of continuity is rich with important insights into Peirce's understanding of cosmic evolution. *Charles S. Peirce's Evolutionary Philosophy* (Cambridge: Cambridge University Press, 1993).

10. These four stages are discussed by Vincent G. Potter, and Paul B. Shields in "Peirce's Definitions of Continuity," *Transactions of the Charles S.*

Peirce Society: A Quarterly Journal in American Philosophy 13.1 (1977): 20–34. N. A. Brian Noble, in "Peirce's Definitions of Continuity and the Concept of Possibility," *Transactions of the Charles S. Peirce Society: A Quarterly Journal in American Philosophy* 25.2 (1989): 149–74, utilizes this distinction to discuss the transition from the second to the third stages.

11. *The Microfilm Edition of the Peirce Papers*, Houghton Library, Harvard University, MS 350, lect. 1.

12. MS 137, p. 13.

13. 4:641.

14. 6:109.

15. 6:109.

16. 6:111.

17. 6:126.

18. 6:126.

19. 7:649.

20. Saint Augustine, *The Confessions of Saint Augustine*, trans. F. J. Sheed (New York: Sheed and Ward, 1943), XI, p. 15. For an interesting and insightful analysis of the way Augustine's entire problematic of time and the paradoxes to which it gives rise involves an inadequate foundationalist ontology, see Gary Calore's "Foundationalism and Temporal Paradox: The Case of Augustine's Confessions," *Anti-Foundationalism Old and New*, ed. Tom Rockmore and Beth Singer (Philadelphia: Temple University Press, 1992).

21. This exclusion was seen to hold both for knife-edged presents and durational presents.

22. 8:86. Italics added.

23. 6:87.

24. 5:284.

25. As Brian Noble points out, "in the earlier stages, Peirce's definition of a continuum suffers from the very same defect that he attributed to Cantor's definition. It implies that a continuum is composed of *individual* instances infinitesimally close together." See Peirce, 7:466 Noble further discusses the way in which Peirce came to recognize that a lapse of time does not consist of a multitude of instants. In a true continuum there are no actual points or instances but rather a number of *possible* points or instants that exceed any nondenumerable multitude ("Peirce's Definitions of Continuity and the Concept of Possibility," pp. 151, 159). Murray Murphey argues that the change in Peirce's conception of possibility was a consequence of his new concept of the continuum (*The Development of Peirce's Philosophy* [Cambridge, Mass.: Harvard University Press, 1961], p. 395), while Noble argues it the other way around (p. 171). Regardless, the interrelation of Peirce's development of possibility and continuity is accepted by both.

26. 6:138.

27. *A Pluralistic Universe*, *The Works of William James*, ed. Frederick Burkhardt (Cambridge, Mass.: Harvard University Press, 1977), p. 154.

28. "The Continuum and the Infinite," *Some Problems of Philosophy*, *The Works of William James*, 1979, p. 88.

29. Ibid., p. 94, n. 17.

30. Ibid., p. 94.
31. Ibid., p. 94, n. 17.
32. Ibid.
33. Ibid.
34. Ibid., p. 95.
35. "Introduction," *Some Problems of Philosophy*, xxxi.
36. 7:651–52.
37. 6:110.
38. 6:203.
39. *Philosophy of the Present*, p. 49.
40. Ibid. Italics added.
41. "The Nature of the Past," *Selected Writings*, p. 350.
42. *The Philosophy of the Act*, pp. 333–35.
43. "The Nature of the Past," p. 349.
44. Ibid., p. 65.
45. The term 'specious present' can be quite misleading. Mead does not mean by this a false or pseudopresent, but rather the temporal spread of the present as opposed to a present characterized in terms of the abstractions of knife-edged moments. A specious present contains within itself past, present, and future.
46. *The Philosophy of the Present*, p. 88.
47. Ibid.
48. Ibid.
49. Ibid., pp. 87–88.
50. Ibid.
51. *The Philosophy of the Act*, pp. 65–66.
52. *Mind, Self, and Society*, p. 351.
53. *The Philosophy of the Present*, p. 13.
54. See chapter 1.
55. *The Philosophy of the Act*, pp. 135–36.
56. Thus, Mead characterizes language as the "field of a fifth dimension" of experience. The first four are, of course, time and the three dimensions of space.
57. *Mind, Self, and Society*, p. 283.
58. This statement is a bit oversimplified, but the interplay of these two perspectives on temporality will be explored in the next chapter.

9. TEMPORALITY AND COSMIC ACTIVITY

1. In previous work I have argued that a process metaphysics can be coaxed out of the collective corpus of the writings of the five major classical American pragmatists (*Speculative Pragmatism*), as well as out of various of them individually in varying degrees, (*Charles Peirce's Pragmatic Pluralism* (Albany: State University of New York Press, 1994); *The Pragmatic A Priori: A Study in the Epistemology of C. I. Lewis* [St. Louis: Warren H. Green, 1976]; *Mead and Merleau-Ponty: Toward a Common Vision*, co-authored

with Patrick Bourgeois [Albany: State University of New York Press, 1992], chapter 3.

2. Mead, *The Philosophy of the Present*, p. 47.

3. Thus, for example, Mead states that "I know of no process that is not that of a living form" (*The Philosophy of the Act*, p. 344). And again, he says of perspectives to which processes give rise, that "the conception of the world without any organisms is one that is without perspectives" (ibid., p. 165). Yet he applies these principles to inorganic situations. See *The Philosophy of the Present*, pp. 52, 65, 86; *Movements of Thought in the Nineteenth Century*, ed. Merritt Moore (Chicago: University of Chicago Press, 1936), p. 317.

4. *Philosophy of the Present*, pp. 14–16.

5. Ibid., pp. 15–16.

6. Ibid., p. 12. Emphasis added; parentheses in text.

7. Ibid., p. 26.

8. Brian John Martine, "Mediation, Immediacy, and Time," *The Journal of Speculative Philosophy* 1 (1987): 270–79.

9. Ibid., p. 276.

10. 6:203.

11. James, *A Pluralistic Universe, The Works of William James*, ed. Frederick Burnhardt (Cambridge, MA and Oxford, England, 1977), p. 116.

12. Ibid., p. 122.

13. Ibid., p. 117.

14. Peirce asserts that the general is that to which the law of the excluded middle does not apply, while the vague is that to which the principle of noncontradiction does not apply (5:448). He then explicitly identifies continuity with generality (8:279; MS 137, pp. 7–12). And, for Peirce, whatever is general or continuous is to some degree vague. Neither the law of noncontradiction nor the law of excluded middle, then, is perfectly applicable to the continuous.

15. Lewis, "A Pragmatic Conception of the A Priori," *Collected Papers*, ed, Goheen and Mothershead (Stanford: Stanford University Press, 1970), p. 232.

16. Neville combines an understanding of concreteness as the fully determinate with an understanding of ontological vagueness as a function of the infinitude of possible harmonies among an infinitude of atomic units. This is of course radically different from the pragmatic understanding of vagueness as a function of the continuity that pervades reality and its temporal dimensions. Though Neville explicitly and creatively draws from Peirce's concept of vagueness in developing his own position, he thus develops it in a quite different direction. See Neville, *Eternity and Time's Flow*, p. 101, for a concise statement of his position.

17. This is intended to deny neither that there emerges an increased richness of spatiotemporal reality as it relates to the human, nor that there are other kinds of reality that emerge within the human world.

18. Dewey, *Experience and Nature*, p. 84.

19. The concrete nature of possibilities as spatiotemporal will be seen in the following discussion. This does not preclude other senses of "possibilities" that are epistemic in nature, as will be discussed later in this chapter.

20. This position is developed in more depth, but with less emphasis on temporality in my *Speculative Pragmatism.*

21. Merleau-Ponty refers to a thickening of time.

22. Dewey, *The Quest for Certainty*, p. 194.

23. James, *A Pluralistic Universe*, *The Works of William James* (1977), p. 144.

24. Ibid., p. 153.

25. Ibid.

26. Ibid., p. 154. See the previous chapter for a somewhat detailed discussion of the role of infinitesimals.

27. The term 'possibility' in the discussion here is being used to cover both 'potentiality' and 'possibility'. Whenever this more general usage occurs, it is because the distinction is not necessary for the discussion and is cumbersome to maintain there.

28. John Dewey, *Experience and Nature*, *The Later Works*, ed. Jo Ann Boydston (Carbondale and Edwardsville: University of Southern Illinois Press, 1981–), vol. 1 (1981), p. 92.

29. This value dimension is developed in some depth in my *Speculative Pragmatism*, chapter 8.

30. Both the terms 'experience' and 'value' are limited to intermeshing matrixes at least one of which is sentient.

31. This sociality of reality, which is here found in the intermeshing of matrixes, was discussed earlier in terms of the sociality of the self in its ongoing temporal dynamics.

32. *Recovery of the Measure*, pp. 50–51. See the previous chapter for the discussion of this point.

33. See especially chapter 4.

34. Here it must be remembered that any selection of a possibility that renders a situation more determinate in some respect, at once opens up new possibilities, thus rendering it less determinate in some respects.

35. The focus here is on human interpretive activity, but even rudimentary animal intelligence, and indeed life itself, embodies the dynamics of a selective response within an environment that it in turn "creatively" transforms.

36. By now it should be clear that there is intended no separation between experience and existence.

37. Lewis, *An Analysis of Knowledge and Valuation* (LaSalle, Ill.: Open Court, 1962), p. 320.

38. Charles Sherover, "Toward Experiential Metaphysics: Radical Temporalism," *New Essays in Metaphysics* (Albany: State University of New York Press, 1987), p. 85. Sherover's way of delimiting potentiality and possibility in terms of temporal modes may grow naturally out of the fact that Heidegger gives a dominant temporal mode to each of the types of human disclosedness, a procedure that, as indicated in chapter 3, may be counterproductive to Heidegger's overall understanding of primordial temporality as a "reaching throughout."

39. This is precisely what leads Bertrand Helm to claim that qualitative possibilities are not in time and to conflate qualitative possibilities with Platonic

forms (*Time and Reality in American Philosophy* [Amherst: University of Massachusetts Press, 1985], pp. 27–28). He makes this specific comment in regards to Peirce's philosophy, and it is discussed by me in some detail in *Charles Peirce's Pragmatic Pluralism*, chapter 5.

40. See especially chapters 4 and 5.

INDEX